WILDLIFE
FACTFINDER

Author
Martin Walters

Editor
Steve Parker

Design
Pentacor

Image Coordination
Ian Paulyn

Production Assistant
Jenni Cozens

Index
Janet de Saulles

Editorial Director
Paula Borton

Design Director
Clare Sleven

Publishing Director
Jim Miles

This edition published by Dempsey Parr, 1999
Dempsey Parr, Queen Street House, 4 Queen Street, Bath, BA1 1HE, UK

Copyright © Dempsey Parr 1999

2 4 6 8 10 9 7 5 3 1

Produced by Miles Kelly Publishing Ltd
Bardfield Centre, Great Bardfield, Essex CM7 4SL

ISBN 1-84084-516-3

Printed in Singapore

WILDLIFE

FACTFINDER

DP
DEMPSEY
PARR

INTRODUCTION

Humans are visual creatures.
We rely on eyesight more than
any other sense—especially to find
out about the world around us from
words and pictures. Lists of facts and
descriptions of events may contain
concentrated knowledge, but adding
illustrations helps to bring the subject
alive. They encourage us to delve
further, appreciate, and enjoy, as well
as to retain the information.

The FACTFINDER series is packed with a huge variety of facts and figures. It also explains processes and events in an easy-to-understand way, with diagrams, photographs, and captions. Fact panels on each main page area contain information for ready reference. Each title is divided into sections that deal with a major aspect of the subject. So look and learn, read and remember— and return again and again.

CONTENTS

Fleas infest the skins of animals, and tapeworms invade their guts. Poisonous scorpions lurk in crevices. Dung beetles burrow into fresh, moist, smelly piles of droppings.

No living thing on Earth survives in isolation. Animals and plants live on, in, under, or near other animals and plants. The world web of wildlife on our planet extends into every nook and cranny. It is never far from us, and affects many aspects of our daily lives.

11

EVOLUTION OF LIFE

O nce upon a time, all life was wildlife. Until humans became so numerous and began to alter the world for their own ends, plants and animals lived as part of nature. Fossils are the remains of animals, plants, and other living things preserved in the rocks. They show the vast variety of wildlife that has existed through millions of years.

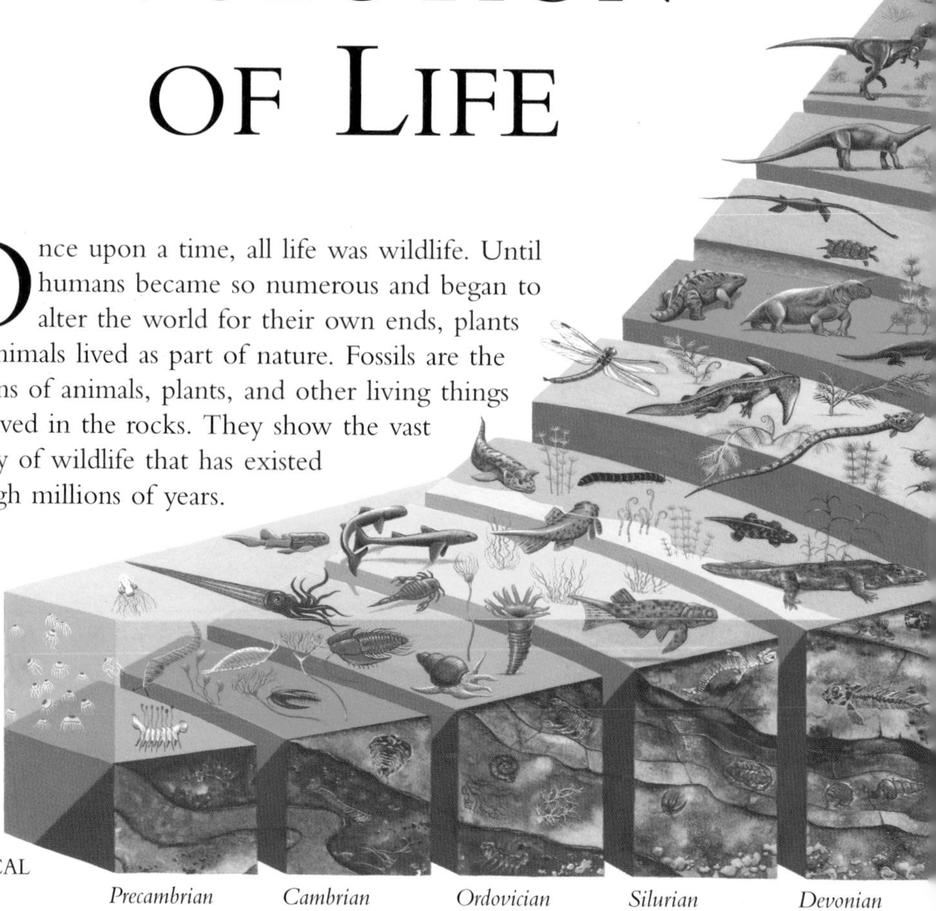

GEOLOGICAL
PERIOD
Millions of years

Precambrian	*Cambrian*	*Ordovician*	*Silurian*	*Devonian*
Long ago to 540	*540 to 505*	*505 to 433*	*433 to 410*	*410 to 360*

Carboniferous
360 to 286

Permian
286 to 245

Triassic
245 to 202

Jurassic
202 to 144

Cretaceous
144 to 65

Tertiary
65 to present

13

STUDY OF WILDLIFE

W̲e could sit and watch wildlife from afar, and simply appreciate its beauty and marvel at its complexity. However if we are to conserve what is left of our wild heritage, we must get much closer to the natural world, and sometimes right inside it.

This involves many branches of the natural sciences. Animal and plant anatomists study the inside structures of living things. Ecologists carry out studies in the field—natural places—

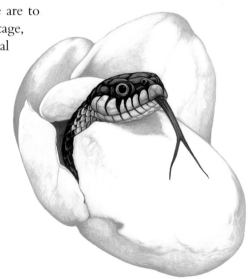

We appreciate herds of great mammals on the African plains, a butterfly flitting through a forest, even a commonplace song bird in the garden. The more we experience wildlife and come to understand it, the greater our pleasure and satisfaction. Books like this one can help to open our eyes to the marvels of nature even in our own backyard. After reading this book, take an opportunity to try the real thing.

INVERTEBRATE ANIMALS

The invertebrates are those animals that have no backbone (vertebral column). In fact, most animals are invertebrates—in all, about 98 out of every 100 of living species. Although they are so numerous, many invertebrates are small, and not as conspicuous as their bigger cousins the vertebrates—the animals with backbones.

The largest main group, or phylum, within the invertebrates is the arthropods. (The name means "jointed-limbed.") This phylum includes the insects (the biggest of any single animal group), the spiders and other arachnids, and the crustaceans, which include crabs, prawns, and shrimps.

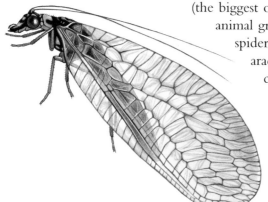

SOFT OR HARD BODIES?
Some invertebrates, such as
jellyfish and worms, have very
soft bodies. Others, however,
notably the arthropods, have
tough, rigid bodies with a shell-like
outer skeleton. Many mollusks, such
as snails, produce a hard chalky shell
to protect the soft body within.

TINY AND SIMPLE ANIMALS

Many invertebrate groups are made up of tiny animals, often with relatively simple bodies. They can only be seen clearly under a hand-lens or even a microscope.

The animals called sponges can be large, but their individual cells are very simple. Indeed, sponges look more like plants than animals. Adult sponges cannot move about, although they do twitch slightly at the surface when touched. They catch their food —even tinier plants and animals—by filtering it out of the water. They also obtain dissolved oxygen in the same way.

Water bears (their scientific name is tardigrades, meaning "slow-moving") are amazing microscopic creatures, with powers as strange as their

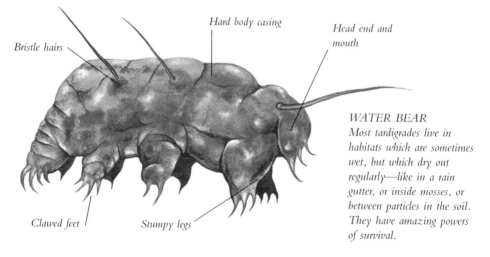

Bristle hairs

Hard body casing

Head end and mouth

Clawed feet

Stumpy legs

WATER BEAR
Most tardigrades live in habitats which are sometimes wet, but which dry out regularly—like in a rain gutter, or inside mosses, or between particles in the soil. They have amazing powers of survival.

appearance. They look like tiny bears, with their plump bodies and stumpy, clawed legs. They are only 0.002 inches (0.05 mm) to just over 0.04 inches (1 mm) in length.

Food being digested

jelly-like body casing

ROTIFER
Rotifers or "wheel-animacules" take their name from the special rings of hair-like structures called cilia on their heads, which look a bit like wheels.

SPONGE
Sponges grow in many shapes, such as cups, tall columns, and flat plates. Most live in the sea but a few dwell in fresh water.

Sponges
10,000 species
- many have spiky skeletons
- adults live fixed to bottom of sea
- body full of holes

Water-bears
400 species
- microscopic
- four pairs of stumpy legs, with bristly claws
- most live in damp places; some in fresh water; a few marine

Wheel-animacules
1,800 species
- microscopic
- crown of cilia (wheel-like in some)
- live in fresh water; some marine; some in damp places

Moss animals
4,000 species
- aquatic; mostly marine
- each animal has a bell-shaped ring of tentacles

23

JELLYFISH AND ANEMONES

This group, the cnidarians or coelenterates, includes the corals and sea anemones as well as all the various kinds of jellyfish. Nearly all these animals live in the sea, and most have circular bodies with a ring of tentacles. Many have barbed cells, which can give a nasty sting! A few are so venomous that they can actually kill people.

Many jellyfish, sea anemones, and corals are beautifully colored, often in shades of pink, red, yellow, and orange. Most have very soft, rather squishy bodies made up of two layers of cells surrounding a jelly-like substance. Some, such as corals, build themselves hard, stony, external skeletons to protect their soft bodies from attack. These external skeletons build up in their millions to form the rocks of a coral reef.

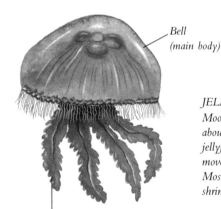

Bell
(main body)

Stinging tentacles capture prey

JELLYFISH
Moon jellyfish (Aurelia aurita) grow to about 16 inches (40 cm) across. Some jellyfish can swim weakly using pulsating movements of the main body or bell. Most catch prey such as small fish, shrimps, prawns, and worms.

MAN-O-WAR
This is actually a colony of small individual jellyfish-like creatures. Its gas-filled float catches the breeze and moves it along, rather like a sailing ship. Its stinging tentacles may be 100 feet (30 m) long.

FRESHWATER TYPES

A few members of the cnidarian group live in fresh water. An example is the hydra. It looks a bit like a miniature sea anemone, with a long stalk and waving tentacles.

COMB JELLY
The comb jellies or ctenophores are similar to jellyfish. They live in the sea and catch tiny animals and plants for food. They also glow at night, making the sea light up with an eerie greenish sheen.

Jellyfish and relatives (Cnidaria/Coelenterata)
10,000 species
• most are marine, some freshwater
• circular body plan
• mouth surrounded by tentacles
• soft-bodied, but some (like certain corals) make hard outer skeletons

Two main subgroups:

Anemones and corals
6,000 species
• marine
• adults attached to a hard surface
• flower-like body shape

Jellyfish and hydroids
4,000 species
• most are marine; a few freshwater
• adults either free-swimming or attached
• some live as colonies
• many have dangerous stings

WORMS

True worms have tube-shaped bodies, divided up into segments or rings. Earthworms are a common kind of segmented worm and are familiar garden creatures. They live in most kinds of soil, especially in old meadows and under grassy lawns. Earthworms have short bristles on their bodies which help them to grip the soil when they are burrowing. An earthworm has no obvious head—the front and tail end look similar, except the head is slightly more pointed.

Most segmented worms live in the sea. These are the bristleworms, which include the familiar lugworms found in seaside sand, and many different kinds of ragworm. Some bristleworms can swim about in the sea, but most of them make burrows or tubes in the sand or mud.

RAGWORM
The flaplike parts along the ragworm's body are called parapodia. They work as gills to absorb oxygen from the water, and as paddles when swimming. The ragworm is a fierce seashore predator and can give a painful bite.

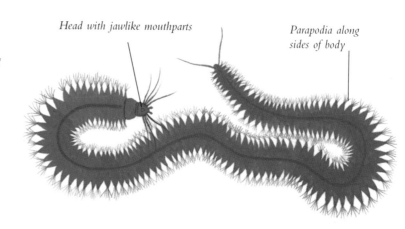

Head with jawlike mouthparts

Parapodia along sides of body

BLOOD SUCKERS

Another group of worms, this time mainly freshwater, are the leeches. Leeches are flattened worms. They have muscular bodies which can contract or extend, and a sucker at each end.

Flattened, leaflike body

Rear sucker

Some leeches prey on worms, mollusks or insect larvae, but many are parasites, sucking the blood of fish or other animals. Some occasionally suck blood from people.

Front sucker under head end

Body compartments or segments

LEECH

A leech can ripple or undulate its body, to swim along. On land it anchors its front sucker, shortens its body, anchors the rear sucker, extends the body, and so on.

Segmented worms

15,000 species
- tube-shaped body
- body divided into segments
- soft body

Three main subgroups:

Earthworms

4,000 species
- most live on land or in fresh water
- hermaphrodite (each individual is both male and female)

Leeches

500 species
- freshwater; some marine or land-living
- some suck blood
- hermaphrodite

Bristleworms

10,500 species
- most are marine
- body has bristles, sometimes used as paddles
- sexes separate
- includes ragworms and lugworms

27

MORE WORMS

Flatworms and roundworms are both common types of worm. Flatworms have bodies that are flattened in outline, almost like a normal worm that has been ironed out. Many are very small, and some live as parasites inside other animals. They show no segments, although some of the parasitic forms have divisions along their bodies.

ROUNDWORMS

About 20,000 different kinds of roundworm or nematode worm have been discovered, but the true number may be nearer to 500,000! Roundworms inhabit just about every habitat, especially soil. But even though they are so common, most of them are very small, only the size of a pinhead. So we do not see them as often as larger worms such as earthworms.

FREE-LIVING
FLATWORMS
These are common under stones in streams and lakes.

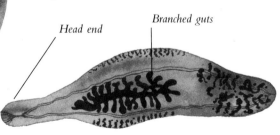

Head end

Branched guts

LIVER FLUKE
This flatworm is a parasite, burrowing into the liver.

ROUNDWORM

The pig nematode worm, Ascaris, lives in the guts and flesh of pigs and similar animals. If pork meat is not cooked properly, this parasite can pass into the human body as tiny eggs, and then grow and cause an infestation. This is why it is important to cook pork meat thoroughly, to kill any roundworm eggs.

Head end

TAPEWORMS

These worms have long, ribbon-like bodies—up to 33 feet (10 m)! Most live as parasites inside the intestines of larger animals, including people. At the front end, the tapeworm secures itself to the gut wall of its host with hooks and suckers.

Proglottis

TAPEWORM

Each section of the body is not a segment, but a separate bag or "proglottis" full of reproductive parts and eggs.

Flatworms
25,000 species
- live in water or damp places, or inside other animals
- soft, flattened body
- no segments
- most are hermaphrodite
- some are free-living, some are parasites
- divided into three groups, the free-living flatworms (Turbellaria), flukes (Trematoda) and tapeworms (Cestoda)

Roundworms
20,000 species
- found in all habitats
- most are free-living; some are parasites
- body rounded, but without segments

Ribbon worms
900 species
- most are marine
- long and ribbon-shaped
- catch and eat other animals

MOLLUSKS

Mollusks are one of the most varied of all the animal groups, containing creatures as different as octopuses, squids, snails, slugs, and shellfish such as oysters and mussels. There are nearly 100,000 species, making them the second largest of all the animal groups, after the arthropods. Even though some mollusks have hard shells, they all have soft bodies. Unlike the arthropods, mollusks do not have jointed limbs. Instead, most mollusks move around on a large flat pad, or foot. The head of a mollusk often has soft tentacles, used to feel the way.

Siphon

MOSTLY MARINE

Most mollusks live in the sea, with only two groups, the bivalves and the snails, having some species that live in fresh water. Some snails have even adapted very well to life

GARDEN SNAIL
Snails, and slugs, including the sea-snails and seaslugs, and the limpets, belong to a group of mollusks known as gastropods, meaning "stomach-foot." This is because they seem to walk or crawl along on their bellies.

on dry land, although they are active only in damp conditions. They seal themselves up inside their shells if the weather gets too dry.

The mollusk group includes the largest of all invertebrates, the giant squid, and also some of the most 'intelligent' invertebrates, octopuses and cuttlefish.

Fleshy body-covering or mantle

OCTOPUS

Squids and octopuses have soft bodies and tentacles with suckers. They have large eyes and keen vision to hunt their prey, which they grab with the tentacles and tear up with the parrotlike "beak" or mouth. They swim by jet propulsion, squirting water from a narrow funnel-like hole or siphon.

Eight suckered tentacles

Mouth in middle of ring of tentacles

Main subgroups of mollusks:

Slugs and snails
77,000 species
• marine, freshwater, and land-living
• most have a single shell, often coiled
• head has tentacles, sometimes tipped by simple eyes
• walk on muscular foot

Bivalves
20,000 species
• most are marine, some freshwater
• hinged shell of two parts (or valves)
• most burrow, or live attached to rocks

Squids and octopuses
650 species
• marine
• tentacles surround the mouth
• large eyes
• hide among rocks or swim in the sea

31

STARFISH AND URCHINS

Starfish belong to a group called the echinoderms, which means "prickly skin," and many of them are indeed rough-skinned or spiny. As well as starfish, the echinoderms include feather stars and sea lilies, brittlestars, sea urchins and sand dollars, and sea cucumbers. All echinoderms live in the sea.

HOW MANY ARMS?
Starfish have arms that spread out from the center, in a starlike pattern. Many kinds of starfish have five arms, although some have seven, and others as many as 14. Underneath each arm there are rows of small flexible tube-shaped feet, which the starfish uses to creep slowly about, or to pry open shellfish to eat.

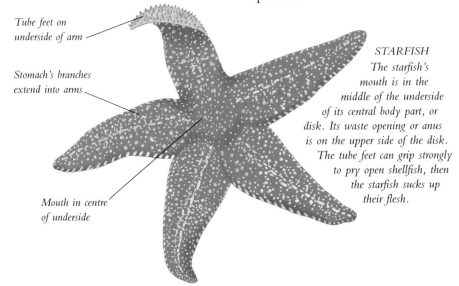

Tube feet on
underside of arm

Stomach's branches
extend into arms

Mouth in centre
of underside

STARFISH
The starfish's
mouth is in the
middle of the underside
of its central body part, or
disk. Its waste opening or anus
is on the upper side of the disk.
The tube feet can grip strongly
to pry open shellfish, then
the starfish sucks up
their flesh.

Sea urchins and sand dollars have rounded bodies, either ball-shaped, or flattened, and many have long spines. They also have a five-part radial body plan, but since they do not have arms, this plan is much less obvious than in starfish. They use their spines to lever themselves along slowly, and can also climb by using their tube-feet. Sea urchins feed by scraping the thin layer of tiny plants and animals, such as corals, off rocks.

Subgroups of echinoderms:

Starfish
1,500 species
• flat and star-shaped
• usually 5 arms (some have more)

Brittle stars
2,000 species
• flat and star-shaped
• usually 5 arms, which are long and brittle

Feather stars and sea lilies
625 species
• either swimming (feather stars) or stalked (sea lilies)
• feed by filtering seawater

Sea urchins
950 species
• body rounded, often hard with long spines
• no arms

Sea cucumbers
1,150 species
• body long and tube-shaped
• tentacles around mouth

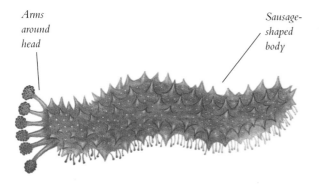

Arms around head

Sausage-shaped body

SEA CUCUMBER
Some sea cucumbers live in shallow water near the shore. But they are far more common at great depths on the ocean floor. They crawl along slowly, sieving the mud and sand for tiny edible particles. In some places, people consider that the sex organs (gonads) of sea cucumbers are a great delicacy to eat.

BUTTERFLIES

Insects have jointed limbs and bodies that are clearly divided into three main sections—the head, thorax (chest), and abdomen. Most adult insects have three pairs of legs, attached to the thorax. Many adult insects also have one or two pairs of wings, also attached to the thorax.

Butterflies and moths, with about 175,000 species, make up the second largest group of insects, after the beetles. They can be found wherever there are trees and flowers,

BUTTERFLY COLORS
The bright colors of the zebra butterfly advertise its presence to potential mates at breeding time.

but are most numerous in tropical and warmer parts of the world. Most species have four large, rather papery and delicate wings, often brightly patterned and colored, covered with tiny scales.

BUTTERFLY OR MOTH?
Most butterflies, like this tropical blue, have clubbed antennae, fly by day, hold their wings out sideways at rest, and are colorful. Most moths have feathery antennae, fly at night, hold their wings together at rest, and have drab colors.

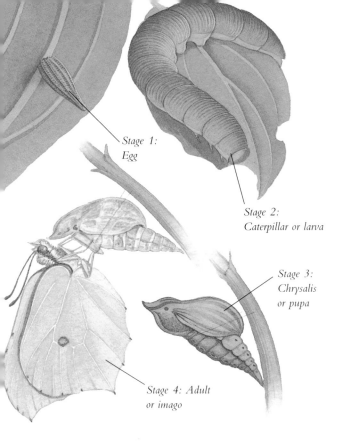

Stage 1:
Egg

Stage 2:
Caterpillar or larva

Stage 3:
Chrysalis
or pupa

Stage 4: Adult
or imago

INSECT LIFE CYCLE

Some insects, such as butterflies and beetles, have a four-stage life cycle (shown above for the brimstone butterfly). They change body shape from one stage to the next, a process known as metamorphosis.

Insects

At least one million species, perhaps ten times that number

- body divided into three main parts—head, thorax, abdomen
- three pairs of legs, in adult and perhaps also immature stages
- most have two pairs of wings, although true flies have one pair
- most are land-living or in fresh water (there are hardly any insects in the sea)

The following pages cover the main subgroups of insects:

Butterflies and moths

175,000 species

- adults have two pairs of wings, often colorful
- adult has long tubelike tongue
- larvae are caterpillars and eat leaves
- most adults feed on nectar from flowers

FLIES, BEES, AND WASPS

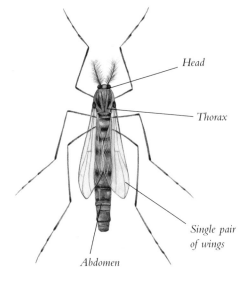

The true flies form a group of more than 90,000 species worldwide. They include houseflies, gnats and midges, craneflies, horseflies, hoverflies, fruitflies, bluebottles, and dungflies. The main feature of the true flies is that they have just one main pair of wings. The second (hind) pair of wings have become thin and club-shaped, used to help the fly balance in flight.

Head

Thorax

Single pair of wings

Abdomen

MOSQUITO

This tiny type of fly is a blood-sucker. It uses its sharp, needle-like mouthparts to drill a tiny hole into the skin, and sucks up blood from its victim. Usually only the female does this, to get nutrients for making her eggs.

MIDGE

Gnats and midges are among the smallest insects. Some feed on flowers or plants. Others are blood-suckers, although few have mouthparts strong enough to pierce human skin.

36

SOCIAL INSECTS

The insect group Hymenoptera includes the familiar honeybees, bumblebees, wasps, and ants, and also more than 100,000 species of parasitic wasps. These are less well known, and many of them are very small.

Some members of this group are social, living together with their own kind in colonies. The members help with different tasks, such as fetching and delivering food, cleaning, and defending the nest from attack. Some ants, called driver ants and army ants, march along in columns when they search for food, and there may be more than 500,000 ants all moving along together.

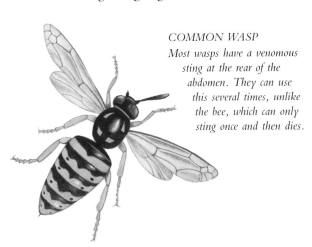

COMMON WASP
Most wasps have a venomous sting at the rear of the abdomen. They can use this several times, unlike the bee, which can only sting once and then dies.

Further subgroups of insects:

Flies
90,000 species
- one pair of transparent wings
- second pair of wings are tiny club-shaped halteres, which work like gyroscopes to give better balance in flight
- larvae are legless grubs or maggots

Bees, wasps and ants
130,000 species
- two pairs of thin wings, the front and rear of each side joined together
- chewing mouthparts
- ants are mainly wingless
- many can sting
- many feed from flowers
- the main types of social insects, along with termites (a different group)

BEETLES AND BUGS

The largest group of insects are the beetles, with more than 400,000 species. In most beetles, the first pair of wings are hard and form a protective case for the wings and body when the insect is at rest. When a beetle flies, it first has to unfurl its thin wings from underneath its hard wing cases. You may have seen a ladybug folding up its wings again just after landing, and tucking them back inside its wing covers.

Weevils form the largest family of beetles, with more than 60,000 species. They feed on plants—often on the seeds, fruits, and flowers. Weevils have a long, curved snout.

LOUSE
Lice are mostly parasites, sucking the blood or body fluids of mammals, birds, and other larger creatures. The louse's legs are shaped like hooks, to cling tightly to the host's skin or hair.

BUGS
People sometimes use the word "bug" to mean any insect, but true bugs or hemipterans are a distinct group of insects. They have beaklike mouthparts, specialized for piercing and sucking. Some, such as assassin bugs and water bugs, feed on other animals, while leaf bugs, cicadas, hoppers, and aphids are mostly plant-feeders.

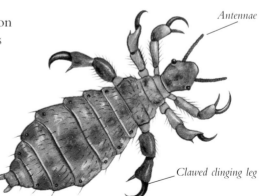

Antennae

Clawed clinging leg

SINGING CICADAS

Cicadas are large, tree-living bugs. The cicada's song is not really a song at all, and comes not from its mouthparts but from the sides of its body. The cicada vibrates a special drumlike part of its hard skin or cuticle, flicking it in and out rather like a lid of a can, only much more rapidly, making a really loud sizzling or fizzing sound. It does this to attract a mate.

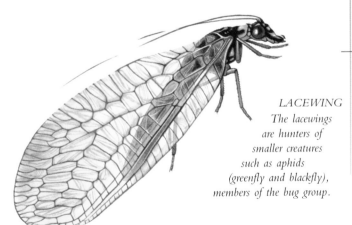

LACEWING
The lacewings are hunters of smaller creatures such as aphids (greenfly and blackfly), members of the bug group.

Further subgroups of insects:

- **Beetles and weevils**
400,000 species
  - front wings are hard wing cases
  - biting mouthparts
- **True bugs**
70,000 species
  - mouthparts for piercing and sucking
  - many have flattened bodies
- **Lice**
3,200 species
  - most are parasites on warm-blooded animals
  - lack wings
  - sucking mouthparts
  - hooklike legs for clinging
- **Lacewings and ant lions**
5,000 species
  - lace-like veins in the wings
  - larvae catch and eat other insects
  - mainly in warm countries

DRAGONFLIES

Dragonflies are some of the most impressive of all insects with their shiny colors and masterful, rapid flight. They can accelerate in seconds to speeds of more than 30 miles per hour (50 km/h), and also hover quite still. These skills allow them to catch other flying insects, such as a midge or gnat, in midair.

Damselflies are smaller cousins of dragonflies. Both dragonflies and damselflies live for a year or longer as larvae or nymphs in streams and rivers. They are powerful predators of smaller animals such as young fish and tadpoles.

EARWIGS

Earwigs may look quite fierce, with their curved claw-like tails, but are actually harmless. They burrow about in the garden soil where they eat a range of plant and animal food, including many harmful grubs.

IN THE WATER
The young or immature mayfly is called a nymph. It lives in fresh water for many months, shedding its skin as it grows, before crawling up a plant stem to change into an adult.

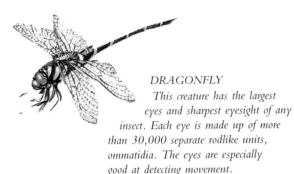

DRAGONFLY
This creature has the largest eyes and sharpest eyesight of any insect. Each eye is made up of more than 30,000 separate rodlike units, ommatidia. The eyes are especially good at detecting movement.

FAST FOOD

Mantises catch and eat other animals, such as large insects and spiders, grabbing them in a flash using their spiny front legs. Their large eyes give them the good vision needed to spot and pounce at their prey.

PRAYING MANTIS
The mantis sits patiently on a plant or flower, waiting until a suitable insect passes close enough. Then it strikes, lightning-quick, to secure its next meal.

Further subgroups of insects:

Dragonflies and damselflies
5,000 species
- mainly large
- two pairs of large, transparent wings
- big eyes
- young (nymphs) are aquatic

Mantises
1,800 species
- triangular head, big eyes
- large front legs with pincerlike claws
- eat mainly other insects
- mainly in warm countries

Earwigs
1,200 species
- flattened body
- pincers at tail end
- very long, flimsy wings

41

GRASSHOPPERS

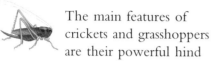

The main features of crickets and grasshoppers are their powerful hind legs, equipped with large muscles. These insects use their legs to make sudden leaps to escape predators. Most members of this group have camouflaged bodies, patterned to blend in with their habitat, and they are very difficult to spot until they jump or fly away.

CHIRPING FOR A MATE

A summer's day in a meadow is seldom without the background chirping sounds of crickets and grasshoppers. These insects have two different methods of "singing." Crickets and katydids rub the veins of their rough front wings together. Grasshoppers and locusts rub their back legs against the front wings. Each leg has a row of hard pegs which make the wing vibrate as they pass over it.

GRASSHOPPER
Most grasshoppers have short antennae, about the length of the head. Many crickets have much longer antennae, sometimes four or five times the body length. Both these types of insects eat plant food such as leaves and shoots.

Large powerful
rear leg

Wings flutter
open to glide
down after leaping

ADULT MAYFLY
The adult mayfly has very small, undeveloped mouthparts. It cannot feed. It lives for less than a day, flitting over water to attract a mate. Then the male dies. The female lasts slightly longer—she lays her eggs and then dies.

A SHORT LIFE
Mayflies live for a year or longer as larvae in streams and rivers. But the adults, which often hatch out all at the same time, live for only a few hours.

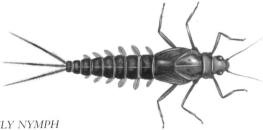

MAYFLY NYMPH
The immature mayfly crawls on the bottom of a pond or stream, feeding on old bits of plant and animal matter.

Further subgroups of insects:

Crickets and grasshoppers
20,500 species
• large hind legs for jumping
• shield-like covering behind the head
• front wings hardened as protection for rear wings

Stick and leaf insects
2,500 species
• leaf insects (mainly Southeast Asia and Australia) look like leaves
• stick insects (tropics worldwide) look like sticks
• mainly in warm countries
• eat various kinds of plant food

Mayflies
2,000 species
• delicate veined wings (usually two pairs)
• three tails
• young (nymphs) are aquatic
• adults very short-lived
• adults stay near fresh water

SPIDERS AND MITES

The arachnids are spiders, scorpions, ticks and mites, and harvestmen. Like insects, arachnids are mostly land-living, with just a few kinds of spiders and mites living in fresh water, and one kind in the sea. With some 86,500 species they make up the second largest group of arthropods, after the insects.

Spiders have a special ability that sets them apart from most other animals. They can make a kind of silken thread. Spiders use their silk to make cocoons for their eggs, as well as for many different designs of traps and webs to catch their prey. Spider silk is one of the strongest materials known. It is stronger than steel wire of the same thickness!

MITES
These small creatures, along with ticks, are tiny eight-legged relatives of spiders.

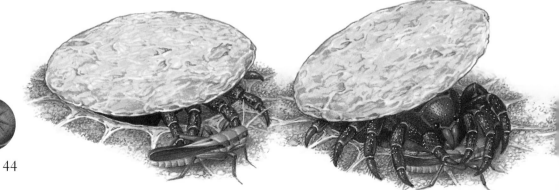

BIRD-EATING
SPIDER
*This type of spider
rarely makes a web. It
runs after its prey and
subdues it with a bite
from the venomous fangs
that all spiders possess.*

TYPES OF SILK

The webs of orb-web spiders are the most
complicated, with two kinds of silk. Thicker,
tough thread is used for the main framework.
The cross-threads are spun from
a more stretchy, sticky type of
silk that traps the victim.

TRAPDOOR SPIDER
*This spider lives in burrows, using
silk to make a covering to the entrance,
with a hinged lid. It sits in its
burrow and leaps out
when a suitable
small creature
passes by. Then
it drags its prey
back down into
the burrow.*

Main subgroups of arachnids:

Spiders
50,000 species
- eight legs
- worldwide, all habitats
- body has obvious waist
  between head part
  (cephalothorax) and
  abdomen
- make webs from silk
- poison fangs

Harvestmen
4,500 species
- like a spider with long,
  stiltlike legs
- rounded one-part body

Ticks and mites
32,000 species
- most are small (mites about
  0.04 inch/1 mm, some ticks
  exceed an inch/2.5cm)
- ticks are parasites of
  mammals, birds or reptiles
- some mites are parasites,
  others free-living
- mites are often brightly
  colored

45

SPIDERS AND SCORPIONS

SUN SPIDER
Not a true spider, the sun or wind
spider has its own arachnid group, the
solifuges. It is extremely fierce and as it
lacks poison fangs, it crushes its prey
to death with its huge
mouthparts.

 Scorpions have long bodies and large pincers that make them look like miniature lobsters. The scorpion's secret weapon is the sting at the end of its long, arching tail. It uses this to immobilize larger prey and to defend itself. Several kinds of scorpion can inflict very painful stings. A few can even kill large mammals, including people.

OUT AT NIGHT

Scorpions are mainly active during the night. By day they hide squeezed into a crevice in a rock, or wedged under a stone or log. Scorpions can feel vibrations in the soil and this is probably how they find their prey. When a scorpion catches up with its prey it grabs the prey with its large pincers, and may then sting the prey to stop its struggling.

MOTHER SCORPION

A female scorpion may lay dozens of eggs. When these hatch, the tiny baby scorpions clamber onto their mother's back. She carries and takes care of them until they can fend for themselves.

MANY LEGS

Centipedes live under logs and among leaf litter. They use their sharp jaws to catch their prey, which they inject with a poisonous bite. Most centipedes can run fast, using their many long legs, one pair per segment. Millipedes are slower moving than centipedes, have more rounded bodies with two pairs of legs per segment, and eat dead and decaying matter.

Centipede

Millipede

Another arachnid subgroup:

Scorpions
1,200 species
- worldwide, in warmer countries
- large pincers
- curved tail with poisonous sting

More groups of arthropods:

Millipedes
8,000 species
- found throughout the world
- mainly in the soil
- herbivores
- many legs
- tubular, rounded bodies
- short feelers

Centipedes
3,000 species
- worldwide
- mainly in the soil or under rocks or wood
- carnivores
- many legs, flattened bodies
- long feelers
- sharp poisonous jaws

47

CRABS AND PRAWNS

Insects are the most common arthropods on land and in the air. But in the water, crustaceans are the most prominent. They include crabs, lobsters, crayfish, shrimps, prawns, barnacles, and water fleas. Most crustaceans live in the sea, but some live in fresh water, and a few, such as woodlice, even live on land, though usually close to water, or in damp habitats.

Tropical prawn

CRABS

There are thousands of kinds of crabs, from giant spider crabs with claws spanning 10 feet (3 m) or more, to tiny species no bigger than a pea. Crabs have very flattened bodies almost entirely protected by their hard shell-like carapace. Crabs walk on four pairs of legs and most have one pair of large curved pincers.

SHORE CRAB
This adaptable crustacean can live in salty or fresh water, and survive in air for several hours. It also eats almost any kind of food, from old seaweed to rotting fish.

SHRIMPS AND PRAWNS

These are like miniature lobsters or crayfish, but they are smaller, lighter, and swim well. They use their small pincers to pick up tiny pieces of food, and their long antennae or feelers to detect objects and also water currents.

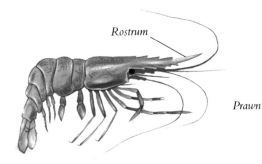

Rostrum

Prawn

SHRIMP OR PRAWN?

Shrimps and prawns are very similar to each other. The main difference is that prawns have a long, pointed extension to the carapace sticking out in front of the head, rather like a beak. This is called the rostrum.

Shrimp

*Swimmerets
(paddle-like swimming limbs)*

Main subgroups of crustaceans:

Crabs
5,700 species
- mostly marine (some in fresh water or on land)
- flattened body, with hard shell
- crawl on seabed
- large pair of pincers
- four pairs of walking legs

Lobsters and crayfish
400 species
- mostly marine (some in fresh water or on land)
- long body
- crawl on seabed
- large pair of pincers
- four pairs of walking legs

Shrimps and prawns
2,000 species
- mainly marine
- long-bodied
- swim and crawl well
- long antennae

SMALL CRUSTACEANS

In addition to familiar prawns and crabs, the crustacean group contains several smaller, less known types of creatures. One of the most beautiful is the fairy shrimp. This looks like a water-dwelling woodlouse and moves along slowly, using wavy movements of its legs to swim. One of the oddest things about the fairy shrimp is that it swims upside-down.

These creatures have no defenses against predators, but they live in temporary pools and puddles where predatory animals, such as fish, cannot survive. When the pool dries out, the adult fairy shrimps die, but the eggs which have gathered in the mud survive until the pool fills again—perhaps years later.

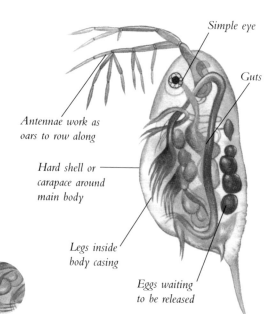

Simple eye

Guts

Antennae work as oars to row along

Hard shell or carapace around main body

Legs inside body casing

Eggs waiting to be released

WATER FLEA

Water fleas (daphnia) have feathery antennae, and they use these to row themselves about in the water. As they move, they breathe using their limbs, which act like gills and take in oxygen from the water. If you look carefully at a water flea you may see the eggs inside its transparent body. The eggs develop inside a special pouch called the brood chamber.

50

BY THE BILLION

The open seas and oceans are the main home of copepods, though some kinds are found in ponds and lakes. Copepods have one long pair of antennae, a hard case around the body, and feathery legs to help gather food. The largest are about as big as a grape; most are tiny.

COPEPODS
These tiny crustaceans are among the most numerous animals on the planet. They occur in shoals of billions and are the basis of many ocean food chains.

PILL-BUG
This crustacean is an isopod, a member of the woodlouse group. It is one of the few dry-land crustaceans and can roll up into a ball for protection.

More subgroups of crustaceans:

Water fleas
480 species
• mainly freshwater, a few marine
• small, often transparent
• bob up and down in the water

Fairy and brine shrimps
175 species
• freshwater pools (brine shrimp in salty water)
• feathery legs for swimming and feeding upside down

Copepods
8,400 species
• mainly marine, some freshwater
• long antennae
• swim jerkily

Barnacles
1,000 species
• marine
• adults have shells and live attached to rocks
• some live inside other animals as parasites

51

FISH

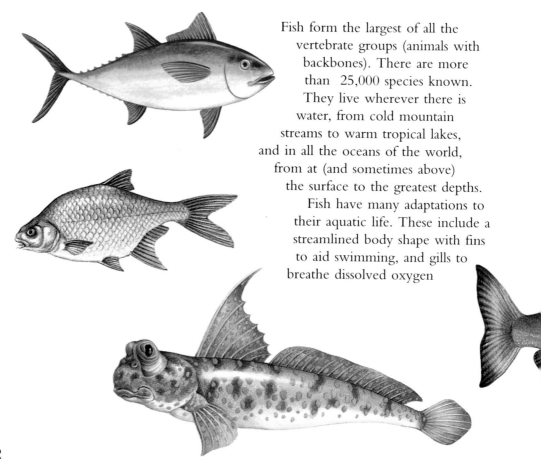

Fish form the largest of all the vertebrate groups (animals with backbones). There are more than 25,000 species known. They live wherever there is water, from cold mountain streams to warm tropical lakes, and in all the oceans of the world, from at (and sometimes above) the surface to the greatest depths. Fish have many adaptations to their aquatic life. These include a streamlined body shape with fins to aid swimming, and gills to breathe dissolved oxygen

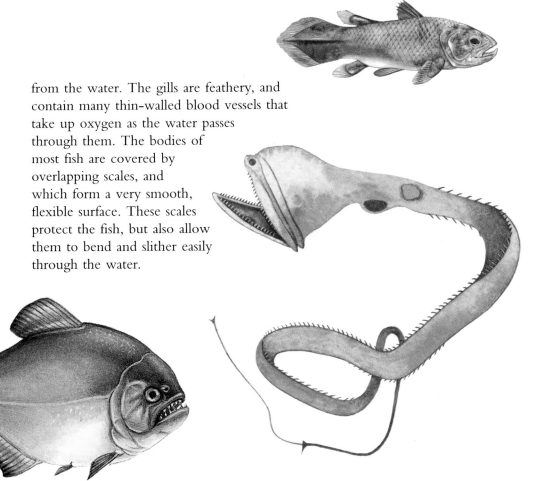

from the water. The gills are feathery, and contain many thin-walled blood vessels that take up oxygen as the water passes through them. The bodies of most fish are covered by overlapping scales, and which form a very smooth, flexible surface. These scales protect the fish, but also allow them to bend and slither easily through the water.

SHARKS AND RAYS

The majority of fish have skeletons made of bones. But the cartilaginous fish have skeletons made of cartilage or gristle. The best known are the sharks. Most are large, fierce predators that use their sharp triangular teeth to kill other animals and tear their flesh. Only a few of the larger sharks are dangerous to humans, but many people are attacked each year, mainly in tropical waters. Dogfish are small, harmless types of sharks.

SKATES AND RAYS
These are like sharks that have been rolled flat! Their side or pectoral fins form large "wings" for flying through the water. Stingrays have sharp spines along their tails and some

Caudal fin or tail

Pelvic fin

can deliver a poisonous sting. Rays and skates often spend long periods resting on the seabed, where they can be difficult to spot—until they move. They eat mainly shellfish, crabs, worms, and similar animals buried in the sand and mud.

Teeth are enlarged versions of the tiny scales (denticles) on the skin

A FEARED MONSTER
One of the most feared sharks is the great white shark (Carcharodon carcharias) which grows to six metres long and can be very aggressive. Its teeth can be 2 inches (5 cm) long. Great whites live in all of the world's warmer oceans, including the Atlantic, Indian, and Pacific Oceans.

Pectoral fin

Cartilaginous fish (Chondrichthyes)
710 species
- cartilage skeleton
- rough scales
- most have sharp teeth
- fins are less flexible than those of other fish

Two main subgroups:

Sharks (Selachii)
370 species
- long and streamlined
- most are active hunters
- all live in the sea

Skates and rays (Batoidea)
340 species
- flattened body with side "wings"
- swim by rippling or flapping their wings
- many eat shellfish or worms in the mud

POND AND LAKE FISH

Most lakes support good populations of fish, unless they are very poor in nutrients, or very badly polluted. Typical pond and lake fish of temperate regions are carp, loach, tench, perch and brown trout. Fish such as carp and tench, which grub about on the muddy pond floor, have small eyes and feelers (barbels) near their mouths, to help them find food.

EVOLUTION IN ACTION

The fish of the East African lakes are world famous, mainly because there are so many closely related species. Lake Victoria (bordered by Uganda, Kenya, and Tanzania) has more than 170 species of cichlid fish, *Haplochromis*. Although they all live in the same lake, the different species tend to eat different prey. Some eat smaller fish and insects, while others

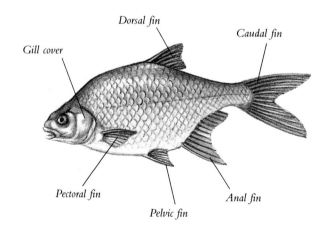

Gill cover

Dorsal fin

Caudal fin

Pectoral fin

Pelvic fin

Anal fin

PERCH
The perch shows clearly the various body parts of a typical bony fish. The pectoral and pelvic fins are paired, one of each pair on each side of the body. The dorsal, anal and caudal fins are unpaired. In bony fish the gills are protected under a bony flap on the side of the head, the gill-cover or operculum.

56

browse on algae. All of these fish probably evolved from one or a few common ancestors. Over millions of years the lake in which they all first lived shrank into several smaller lakes, which then expanded and merged again. These changing conditions led to the evolution of new and different species, each specializing on different food, but all with similar body features.

More groups of fish:

Jawless fish (Agnatha)

70 species

- simple cartilage skeleton
- no jaws
- no scales
- no proper fins
- many pairs of gills
- includes lampreys and hagfish

Lungfish (Choanichthyes)

6 species

- leglike fleshy fins
- lungs as well as gills
- survive dry periods by burying into lakebed mud

Bony fish (Osteichthyes)

24,500 species

- the main fish group
- bone skeleton
- smooth scales
- live in all watery habitats

PIKE
The pike is an ambusher. It prefers weedy ponds and lakes, also reservoirs and slow-flowing rivers all around the Northern Hemisphere. It lurks in water plants and dashes out to grab victims with its huge mouth.

RIVER FISH

Fish such as trout and salmon spend some of their lives in fast-flowing rivers. These fish must swim constantly to maintain their position in the river, or to make progress upstream. They have very muscular and flexible bodies and can even leap over small weirs or waterfalls.

Slower-moving rivers in temperate regions are home to fish such as catfish, dace, and roach. Barbel and gudgeon are also river fish. Tropical streams and rivers are home to hundreds of fish species. Many attractive freshwater tropical fish bred for aquaria, such as the bright iridescent red-and-blue neon tetra, come from the Amazon region of South America.

GAR PIKE
Slim and streamlined, the several kinds of gars feed by slashing at smaller fish with their long snout.

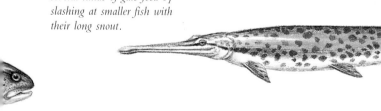

58

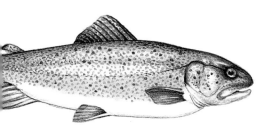

FOOD FISH
Some river fish, such as the rainbow trout, are bred specially for food. This species came originally from western North America, but as a result of fish farming it has now spread to most parts of the world.

PIRANHA
The typical piranha is quite small, rarely growing larger than a human hand. One or two piranha on their own are unlikely to attack large animals. But as soon as a shoal senses blood in the water, the fish attack in a feeding frenzy.

AN INFAMOUS FISH

Another famous, or rather infamous, Amazon river fish is the piranha (*Serrasalmus*). Piranhas have sharp teeth and hunt in big shoals or packs. Working together, they can strip the flesh from creatures which fall into the river. Some kinds of piranha are vegetarian, however, feeding on nuts and fruit.

Fish records

Smallest fish
The tiny Philippine Goby (*Pandaka pygmaea*) at half an inch (12 mm) total length.

Biggest fish
The enormous whale shark (*Rhincodon typus*) at 50 feet (50 m) long. Like all sharks, this is a cartilaginous fish.

Biggest bony fish
Probably the beluga (*Huso huso*), a kind of sturgeon found in the Caspian Sea, which reaches 12 feet (3.5 m) in length. Another very big bony fish is the South American arapaima (*Arapaima gigas*), which grows to 15 feet (4.5 m).

Fastest fish
The sailfish, swordfish, and mako shark have all been timed at over 50 miles per hour (80 km/h).

SEASHORE AND REEF FISH

The fish of coastal habitats must cope with ever-changing conditions. Tides rise and fall, waves batter the shore, fresh water (as rain) dilutes salt water, and small rock pools can ice over in winter or become as warm as a bath in summer.

CLINGING ON FOR LIFE
Gobies and blennies are common fish of rocky shores in temperate seas.

They cling on tight with their spiny fins or hide in crevices when the waves break. Mudskippers are common in mangrove swamps in West Africa, Southeast Asia and Australia. They spend much of their time partly submerged in mud or shallow water, but they can climb out into the air – and even clamber up mangrove stems using their arm-like fins.

MUDSKIPPER
The mudskipper has large gill chambers which can store water, allowing the fish to 'breathe' while in air. The dorsal fin is raised and lowered as a territorial signal.

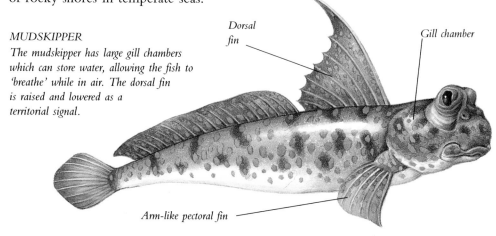

Dorsal fin

Gill chamber

Arm-like pectoral fin

ON THE REEF

Coral reef communities teem with shoals of brightly colored fish. They use the many crevices in the coral rocks to hide from predators. One of the most remarkable of all tropical marine fish is the seahorse, named from its resemblance to a tiny horse. Some species have tasseled fins, making them blend into the background of frilly seaweeds and corals.

PUFFERFISH
By swallowing water, this fish becomes too big for other fish to swallow.

SHOALING
Small fish dart and turn together in a vast shoal, making it difficult for a predator to single out a victim.

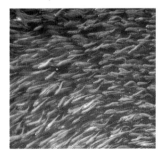

Fish of the seashore
Gobies
Blennies
Wrasses
Butterfish
Mudskippers

Coral reef fish
Parrotfish
Groupers
Clownfish
Moorish idols
Seahorses
Stonefish

Fascinating fish
• Some clownfish live among the stinging tentacles of sea anemones. The fish themselves are protected by a special layer of body slime or mucus. Any other kind of fish risks being stung to death and eaten by the anemone.
• The stonefish is one of the world's most poisonous fish. It has sharp spines which can inject a deadly venom.

OCEAN FISH

 Many fish live in the open sea. Some are adapted to life near the surface. Here, where the sunlight penetrates, there is plenty of food, from tiny algae and microscopic plankton, through to shrimps, squid and baby fish. Surface-dwelling ocean fish tend to be fast swimmers with keen eyesight, such as mackerel, tuna and marlin.

THE BOTTOM OF THE WORLD

The depths of the ocean are cold and dark, but the mud of the ocean floor offers rich pickings of worms and other food. Deep-sea fish include the bizarre gulper eel and the

MARLIN
This sleek billfish folds its fins against its body and goes at full speed as it thrashes its tail.

Muscular body

Crescent-shaped tail for fast swimming

BLUE-FIN TUNA
The tuna or tunny hunts smaller fish and squid. It is itself hunted by larger fish such as mako and tiger sharks, as well as our own fishing fleets.

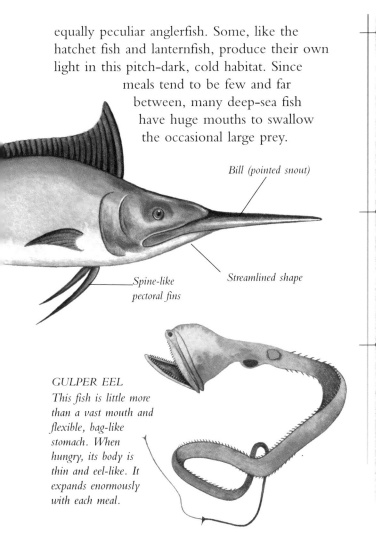

equally peculiar anglerfish. Some, like the hatchet fish and lanternfish, produce their own light in this pitch-dark, cold habitat. Since meals tend to be few and far between, many deep-sea fish have huge mouths to swallow the occasional large prey.

Bill (pointed snout)

Spine-like pectoral fins

Streamlined shape

GULPER EEL
This fish is little more than a vast mouth and flexible, bag-like stomach. When hungry, its body is thin and eel-like. It expands enormously with each meal.

Fish of the upper ocean
Marlin
Swordfish
Tuna
Mackerel
Herring
Flying fish
Wolf-fish
Ocean sunfish
Plaice

Fish of the deep ocean
Halibut
Gulper eel
Anglerfish
Ribbonfish
Tripod fish
Hatchet fish

Fish in mid air
• Flying fish can leap as high as 20 feet (6 m) above the sea's surface, then use their wing-like fins to glide for up to 1,000 feet (300 m). They are found mainly in tropical seas.

63

AMPHIBIANS

The amphibians group includes frogs and toads, newts, and salamanders, and the strange, worm-like caecilians. Many seem slow, peaceful creatures. But they are all deadly predators, hunting live prey — from tiny gnats and flies, to rats, snakes and even bats.

A typical amphibian has soft, damp skin and lives in or around water, or in damp places. Like fish, amphibians are cold-blooded animals, which means they take on the temperature of their surroundings. So they are active only in warm conditions. In cold weather they hide in mud or soil, under stones or among damp logs.

TWO LIVES

Although many adult amphibians can live on land, even in deserts, most breed in water. Indeed, the name 'amphibian' means 'two lives', the first part in water and the second on land. The jelly-covered eggs, known as spawn, hatch out as the familiar tadpoles (larvae). These live in water and breathe by gills, before developing lungs, changing shape, and growing into adults. This process of changing body shape while growing up is known as metamorphosis.

Amphibians can be found wherever there is fresh water. Most species live in warmer places such as tropical rain forests. Their main defense is to hide, or to rely on the poison glands in their skin.

FROGS AND TOADS

The distinction between frog and toad is not clear-cut. As a general guide, those with drier, warty, dull-colored skin and tubby bodies, who tend to walk or waddle, are usually toads. Those with moist skin and slim, brighter-colored bodies, who leap rather than walk, are usually frogs.

FROG OR TOAD?

The common frogs and toads of Europe fit these descriptions. But elsewhere, especially in the tropics, there is a huge variety of shape and size. Many tropical frogs are brightly colored, warning that their skin is very poisonous, while others are almost transparent.

The flying frog has broad webbing on its feet, which it uses to glide from tree to tree. Tree frogs are green to match their leafy surroundings and they cling on with their sucker-like toes, hopefully unnoticed. The spadefoot toad uses its powerful rear legs to dig a hole in the ground, where it can hide in safety.

parachute-like webbed toes

MARINE TOAD
Like other amphibians, this toad takes only moving prey, from worms and beetles to flies and fish. If an animal stays still the toad ignores it.

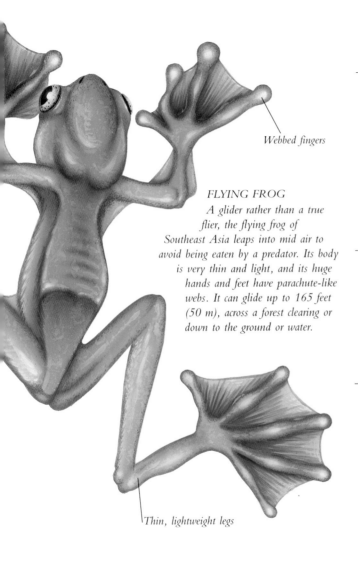

Webbed fingers

FLYING FROG
*A glider rather than a true
flier, the flying frog of
Southeast Asia leaps into mid air to
avoid being eaten by a predator. Its body
is very thin and light, and its huge
hands and feet have parachute-like
webs. It can glide up to 165 feet
(50 m), across a forest clearing or
down to the ground or water.*

Thin, lightweight legs

Amphibians (Amphibia)
4,000 species
- cold-blooded
- live in water or damp places
- most breed in water
- skin not waterproof
- breathe through skin, gills or lungs

Main groups of amphibians include:

Frogs and toads (Anura)
3,500 species
- jelly-like eggs
- aquatic larvae (tadpoles)
- adults have long hind legs
- most have webbed feet
- no tail when adult

Frog and toad records
- The smallest frogs or toads are less than an inch (2.5 cm) long.
- One of the largest is the marine (giant or cane) toad, growing to over 9 inches (23 cm) in head-body length.

SALAMANDERS AND NEWTS

Newts and salamanders have similar lifestyles to frogs and toads. However, their bodies are a different shape, longer and slimmer, with four legs of roughly equal size—and a long tail. They are rarely seen in the wild as they tend to lurk in damp crevices, or hide among water plants. They walk or waddle rather clumsily, but most can swim well with a fishlike bendy motion. Newts have tails that are flattened to give them extra thrust when swimming.

LUNGLESS SALAMANDERS
The largest group of salamanders is the lungless salamanders. They are common in streams and damp woods in parts of North America. As they have no gills or lungs, these amphibians breathe entirely through their skin. So they must keep their

FIRE SALAMANDER
This salamander gets its name from its supposed ability to withstand flames and fire. In fact, when people gathered damp wood for a log fire, the wood sometimes had a salamander hiding in it. This creature soon felt the heat—and seemed to emerge miraculously from the flames.

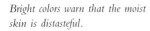

Bright colors warn that the moist skin is distasteful.

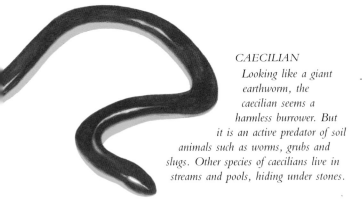

CAECILIAN
Looking like a giant earthworm, the caecilian seems a harmless burrower. But it is an active predator of soil animals such as worms, grubs and slugs. Other species of caecilians live in streams and pools, hiding under stones.

skin damp, which allows oxygen to pass through it. If they dry out, they suffocate and die. Other groups include the burrowing mole-salamanders of North America, the almost legless congo eels and sirens, and mudpuppies.

MUDPUPPY
An inhabitant of lakes and rivers in eastern North America, this salamander breathes using its feathery gills and eats small fish, water snails, crayfish and freshwater insects.

More groups of amphibians:

Salamanders, newts and mudpuppies (Urodela)
360 species
- long, flexible body
- long tail
- small legs
- larval or tadpole stage (as in frogs and toads)

Caecilians (Apoda)
163 species
- legless
- worm-shaped
- almost blind
- live in damp soil of tropical forests

Biggest amphibian
The largest of all amphibians are the Asian giant salamanders of cold streams in China and Japan. They can grow to more than 5 feet (150 cm) in total length and weigh 90 pounds (40 kg).

REPTILES

Dry, scaly skins characterize the reptiles, which are well adapted to life on land, though some are equally at home in the water. Many reptiles are found in dry habitats such as deserts, but some, notably the turtles, are almost entirely aquatic.

Most reptiles lay eggs. But some lizards and snakes give birth to young or babies, which resemble miniature adults. Even the aquatic reptiles return to dry land to lay their eggs. Reptile eggs are soft and leathery, not hard-shelled like the eggs of birds.

IN COLD BLOOD

Reptiles are cold-blooded and many need to bask in the sun to get warm enough to move quickly. This means there are no reptiles in the coldest parts of the world, on mountaintops and near the poles.

Lizards and snakes are the most familiar reptiles, but the group also includes turtles and crocodiles, and the strange tuatara. Most reptiles creep, slither, or crawl along. A few, however, such as the racerunners and water dragons, can run fast using just their hind legs. Some, such as the basilisk, can even run across the surface of the water.

TURTLES

Loggerhead turtle

Turtles, terrapins, and tortoises make up the reptile group called the chelonians. Most of the members live either in fresh water or in the sea. The sea turtles are large, and are surprisingly graceful swimmers. The females come ashore on favored beaches to lay their eggs in pits dug in the sand. The eggs incubate in the heat of the sun and hatch weeks later. The tiny young must race to the sea, usually under cover of darkness, to avoid being eaten by gulls, lizards and other animals who gather for a feast.

FRESHWATER CHELONIANS

Many different kinds of turtles and terrapins are found in rivers, ponds, and swamps, mainly in the warmer

MARINE TURTLES

The fully sea-dwelling marine turtles include the leatherback (largest of the group at more than half a ton in weight), loggerhead, hawksbill, and Ridley turtles. Most cannot withdraw their head and legs into their shell, and rarely come ashore except to bask in the sun or lay eggs.

Green turtle

Rear legs used for steering

Front legs are powerful flippers for swimming

regions of the world. Many have long, flexible necks to enable them to reach the surface and breathe easily while the rest of the body remains safely submerged. The snapping turtles are fierce predators with powerful jaws. They can catch fish, frogs, young turtles, and even small birds. (Tortoises are shown on the next page.)

FRESHWATER TURTLE
The Murray River turtle is a side-neck, hiding its head in its shell by bending its neck sideways.

TORTOISES

Tortoises are land-dwelling members of the turtle group, the chelonians. They are rather slow-moving reptiles, found mainly in warm, dry regions. They use their hard, horny jaws to tear at grasses, fruit, and leaves. Although largely vegetarian, tortoises occasionally eat small animals such as insects and slugs.

TORTOISE SHELL

The shell of a tortoise is very hard and forms effective protection against all but the most ingenious of predators. When danger threatens, the tortoise simply pulls in its limbs, head and tail and waits until the coast is clear.

Tortoises were once popular as garden pets, especially in parts of Europe. But they were imported

DESERT TORTOISE

Many tortoises are colored dull brown or green, to match their surroundings and be camouflaged from predators. The desert tortoise is an excellent digger and excavates a long burrow, where it hides from the intense daytime heat. It emerges in the cool night to feed.

Upper part of shell is carapace

Lower part of shell is plastron

Shell covered with horny plates called scutes

Leopard tortoise

from warmer countries and were not adapted to the cold winters of central and northern Europe. Wildlife laws now forbid the importing of tortoises in most regions. This saves them from suffering and dying during the cold winter weather.

GIANT TORTOISE
There are several kinds of giant tortoises, each living on an island or group of islands. They eat various types of plant food.

More groups of reptiles:

Tortoises (subgroup of Chelonia: Testudinidae)
40 species
- hard shell
- tough, clawed feet lack the webs of turtles and terrapins
- live mainly on dry land, even in deserts

The biggest tortoises
Giant tortoises live on certain isolated islands, especially in the Atlantic and Pacific Oceans.
- The giant tortoise of the Galapagos Islands in the Pacific grows to about 4 feet (1.2 m) long and weighs up to 500 pounds (225 kg).
- Even larger is the giant tortoise of the island of Aldabra off East Africa, which can reach 6 feet (1.8 m) in length.
- Giant tortoises can live for 150 years or more.

BIG LIZARDS

Most species of lizard are small or medium-sized. However, certain groups contain large species that can be quite impressive and dangerous to humans.

The iguana group has several large species. One of the most famous is the marine iguana of the Galapagos Islands in the Pacific, off the coast of Ecuador. This species grows to 70 inches (1.75 m) in length, and as its name suggests, it spends much of its time in the sea, eating seaweeds. It is the only truly marine lizard. It can dive deep below the surface and stay submerged for 20 minutes at a time.

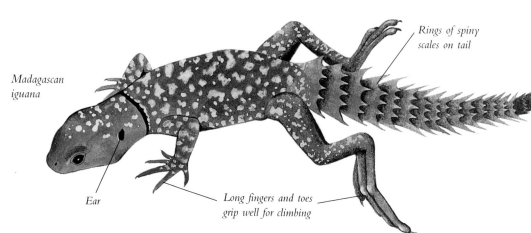

Rings of spiny scales on tail

Madagascan iguana

Ear

Long fingers and toes grip well for climbing

THE ORIGINAL DRAGON?

The Komodo dragon is the world's largest lizard. It is a type of monitor lizard and may explain legends about dragons. It can grow to more than 10 feet (3 m) long and 220 pounds (100 kg) in weight. It feeds on other animals, including pigs and small deer.

FRILLED LIZARD
This formidable lizard usually keeps its frill folded against its neck. If threatened, it extends the frill like a multicolored fan and hisses loudly with its mouth wide open, to scare away the attacker.

COMMON IGUANA
This American lizard is bright green when young. The bands on the body and tail darken with age.

More subgroups of reptiles:

Lizards (Sauria or Lacertilia)
3,750 species
- Live in most habitats, including deserts, trees, swamps, rivers, and the sea
- Most have four slim legs, long tail
- Some lizards, like slow-worms and certain skinks, are legless
- Most are fast and active when warm

Main groups of large lizards:

Iguanas (Iguanidae)
650 species
Monitor lizards (Varanidae)
37 species
Beaded lizards (Helodermatidae)
2 species
- Includes the gila monster and Mexican beaded lizard, the only two lizard species with poisonous bites

77

SMALLER LIZARDS

Most lizards prefer dry habitats and are active during the day, when the sun warms their bodies. The group with the most species is the skinks. These are rather fat-bodied lizards, often found in sandy habitats. Another large group is the geckos. These are unusual because they hunt during the night. They are mostly found in tropical climates that are warm at night as well as by day. Geckos have very large eyes to help them find their food even in the dusk.

WALKING UP WALLS

Most geckos have another special adaptation. They can run up vertical rocks, walls, or even windows and across ceilings. Their toes have ridges with millions of tiny hooks to help them to grip the smoothest surface.

Most lizards have large eyes and good vision

Strong legs for running and climbing

BANDED AGAMA
The bright colors of many lizards are for courtship, to attract a mate at breeding time. The colors also help the lizard to blend in with the bright flowers and plants in its tropical home.

The thorny devil or moloch is covered in spikes. These protect it from attack and also provide a surface that collects droplets of dew, for the lizard to drink in its Australian desert home.

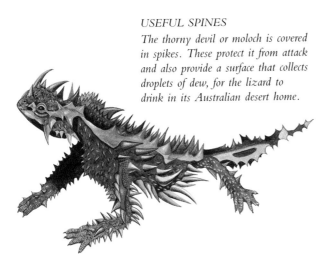

LEGLESS LIZARDS

Slowworms are small, legless lizards. In fact they look more like snakes than worms, but unlike a snake, a slowworm has eyelids.

Bright blue tongue

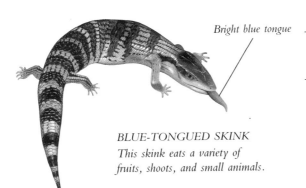

BLUE-TONGUED SKINK
This skink eats a variety of fruits, shoots, and small animals.

Main groups of smaller lizards:

- **Skinks (Scincidae)**
  1,275 species
- **Geckos (Gekkonidae)**
  800 species
- **Chisel-teeth lizards (Agamidae)**
  300 species
- **Wall and sand lizards (Lacertidae)**
  200 species
- **Whiptails and racerunners (Teiidae)**
  225 species
- **Slowworms and relatives (Anguidae)**
  75 species
- **Chameleons (Chamaeleontidae)**
  85 species
- **Girdle-tailed lizards (Cordylidae)**
  50 species
- **Snake lizards (Pygopodidae)**
  30 species

79

NONPOISONOUS SNAKES

PYTHON
The reticulated python, from
Southeast Asia, can kill
mammals as large as a pig
or a small deer using the
constrictor method.

Body thrown
into S-curves
as snake slithers

Although
we may
think of snakes
as dangerous and
unpleasant, in fact
only about one fifth
of all snake species are
poisonous. And only about a hundred of these
are aggressive enough, with teeth strong enough,
and venom powerful enough, to harm people.

SQUEEZED TO DEATH

Nevertheless, some of the larger nonpoisonous snakes
sometimes pose a threat—particularly those that kill
their prey by suffocation, such as the larger pythons and
boas. These snakes are called constrictors. They wrap their
muscular bodies around their prey and slowly squeeze it to

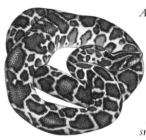

ANACONDA
The largest snakes are the reticulated python of Southeast Asia and the anaconda of South America, which may reach 33 feet (10 m) in length. The amethystine python is Australia's largest snake; it can grow as long as 23 feet (7 m).

Neck and gullet are elastic to stretch for swallowing large victims

Eyes lack eyelids, giving snakes a "glassy" stare

Large scales on underside tilt to grip the ground

death, before devouring it whole. Each time the victim tries to breathe, the snake grips tighter until breathing becomes impossible. After a large meal, these snakes rest for weeks as they digest the victim. Pythons often rest coiled in tree branches.

Another reptile subgroup:
Snakes (Serpentes or Ophidia)
2,400 species
• long, narrow body
• no legs
• no eyelids
• no eardrums
• all snakes hunt prey, a few specialize on scavenging or items such as eggs

The main (mostly non-poisonous) snake groups include:

Typical snakes (Colubridae)
1,500 species
Blind snakes (Typhlopidae)
160 species
Boas (Boidae)
40 species
Pythons (Pythonidae)
30 species

POISONOUS SNAKES

Poisonous snakes use their long teeth or fangs to inject venom into their prey when they bite. Although some snakes are aggressive, most people get bitten when they tread on a poisonous snake by accident, and the snake strikes back in self-defense. Snake bites still kill thousands of people each year, although antivenoms are now usually available for treatment.

FRONT-FANGED SNAKES

Some of the most dangerous snakes, such as cobras, mambas, and the taipan of Australia, belong to the front-fanged snake family. Although their fangs are not as efficient at delivering venom as those of the other main group, the vipers, the cobras, and coral

Hood spread to reveal eyespots

CAPE COBRA
Cobras can extend the loose ribs on either side of the head and neck, to form a hoodlike flap of skin. This warns that the snake is ready to strike. It also raises itself off the ground to get a clearer view and easier aim. The king cobra preys mainly on other snakes and also lizards. The female Indian cobra is unusual among snakes since she guards her eggs until they hatch.

Flicking tongue detects airborne scents

TAIPAN
This sleek, fast, rapid-striking snake lives in north-east Australia and New Guinea.

snakes often hang onto their victim and make chewing movements to increase the flow of poison. The sea snakes include some of the world's most venomous species. Although sea snakes have highly toxic venom, they are not usually aggressive and tend not to inject much venom when they do bite.

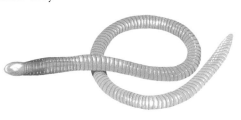

WORM-LIZARDS
These reptiles have their own subgroup, the amphisbaenids. They look like large earthworms and burrow in the soils of tropical forests, hunting small animals.

Groups of poisonous snakes:

Front-fanged snakes (Elapidae)
240 species
• includes death adder, cobras, mambas, taipan, kraits, coral snakes, tiger snakes, and sea snakes

Vipers (Viperidae)
190 species
• includes adders, vipers, cottonmouth, fer-de-lance, bushmaster, copperhead, rattlesnakes, sidewinder

• Spitting cobras from Africa and Southeast Asia sometimes spray venom into the eyes of a victim.
• The largest poisonous snake is the king cobra, over five metres in length.
• The most dangerous is probably the Indian cobra, which kills thousands of people each year. The snakes with the deadliest venom are the black mamba of Africa and the taipan of Australia.

CROCODILES

Crocodiles, alligators, caimans, and gharials form the crocodilian group of reptiles. They are the largest of all the reptiles, and also the most similar to their ancient cousins the dinosaurs, with their heavy skin and powerful jaws. Their jaws are very muscular and can clamp shut with huge force. They are lined with sharp teeth with which they can easily tear the flesh of their prey—mainly fish, mammals, and water birds. The gharial is a specialist fish-eater from Asia. It has a more delicate, thinner snout.

HUNTING METHODS

Most crocodilians lie in wait for their prey, almost submerged. Then, when the prey is within range they surge

Eyes on top of head

ESTUARINE CROCODILE

This huge reptile an be found in the open sea as well as in rivers and estuaries. It is a powerful swimmer, and large and strong enough to overcome most victims, from deer to sea turtles and large fish. Because of persecution and hunting, it is now rare, and is protected by law in most regions.

Nile crocodile

forward and drag it under water, holding it down until it drowns. One of the strangest things about crocodiles and alligators is that they cover their eggs with rotting compost. This keeps the eggs warm, so that they can incubate and the babies develop inside.

Flattened tail for swimming

CAREFUL MOTHER
The mother crocodile guards her eggs until they hatch. The babies make squeaking noises and she carries them carefully to the water in her mouth.

The crocodilian group of reptiles has three main subgroups:

Crocodiles (Crocodylidae)
14 species
Alligators and caimans (Alligatoridae)
7 species
Gharial (Gavialidae)
1 species

Largest reptiles
These are the estuarine or saltwater crocodiles of Southern Asia and Northern Australia.

- These giants can reach over 25 feet (7.5 m) in length.
- They feed mainly on fish and crustaceans, but will also take other vertebrates, and can be a danger to people.

BIRDS

Active mainly by day, conspicuous, and often colorful and noisy, birds are well known to everyone. Many birds announce their presence by loud calls or musical songs. There are some 9,000 kinds or species of birds around the world, in all land habitats, and also on water and at sea.

Birds are the most prominent animals in many of the world's ecosystems, from the polar regions to the tropics. This is partly because their powers of flight gives them extra mobility. Even the Arctic wastes thrill to birdsong during the busy breeding season.

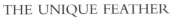

THE UNIQUE FEATHER

Birds are warm-blooded and insulated by their feathers, which are their unique feature. So birds can be active even under cool or cold conditions.

As well as keeping birds warm, feathers play a vital part in flying. The main flight feathers rest in sockets in the wing bones. The number of feathers a bird has depends on its species, size, and sex.

BIRDS EVERYWHERE

Since birds lay hard-shelled eggs they do not need to depend on water for breeding, and this has helped birds to colonize most regions, even isolated oceanic islands.

FLIGHTLESS BIRDS

The best-known flightless birds are probably the ostrich, which lives on the African plains, and the emu of Australia. The cassowary is another flightless bird from Australia, while the rhea dwells in South America. All of these birds can run fast and use their powerful legs and bills (beaks) against predators.

The chicken-sized kiwis of New Zealand are another flightless group. They have fluffy, furlike feathers and tiny wings.

FLYING UNDER WATER
The largest group of flightless birds is the penguins, from the Southern Hemisphere. Their wings are like tough paddles. Penguins flap their wings when submerged to achieve a sort of underwater flight.

LOST THE POWER OF FLIGHT

In addition to these groups of flightless birds, there are some members of flying groups which have lost the power of flight. Flightless rails are found on certain isolated islands. New Zealand has a flightless parrot, the kakapo, which hides under rocks and undergrowth by day and emerges at night to feed.

Wings

Male ostrich

EMPEROR PENGUINS
Most penguins live on rocky islands, the icy shores of mainland Antarctica, or on drifting icebergs and ice floes. They catch fish, squid, shrimps, krill, and other sea animals for food. The emperor is the largest penguin, about 4 feet (120 cm) tall.

Birds (Aves)
9,000 species
• possess feathers
• lay hard-shelled eggs
• scaly legs
• front limbs are wings
• beak without teeth
• warm-blooded

Flightless birds include:
Ostriches
1 species
• Africa
Emus
1 species
• Australia
Rheas
2 species
• South America
Cassowaries
3 species
• Australia, New Guinea
Kiwis
3 species
• New Zealand
Penguins
16 species
• Antarctic region and southern oceans

SEA BIRDS

Many different kinds of bird have become adapted to life at sea. Best known are the gulls and terns, hardly ever absent from a visit to the coast or on a boat trip. Gulls are equally at home far inland, and often feed on open fields, especially in the winter. Terns are graceful sea birds, which catch their fish prey by darting, near-vertical dives.

The real expert oceangoers are the albatrosses and petrels, which spend weeks far out at sea. Albatrosses save energy by gliding on upcurrents from the waves as they cross the oceans. Very different in build, but nevertheless well adapted for life at sea are the auks. Their plump bodies and tight feathering insulate them well, and they are also adept divers.

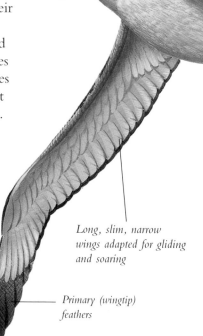

Wandering albatross

Long, slim, narrow wings adapted for gliding and soaring

Primary (wingtip) feathers

STORM PETREL
The only time this bird alights on land is during the breeding season, on remote islands or outcrops.

90

BIG BILLS

The puffin is an auk and can hold several fish at once in its strange bill. Cormorants and pelicans are related to the puffin and share the feature of a large bill, equipped in the case of the pelican with a baggy storage pouch.

Colorful sharp-edged bill grasps slippery food

Legs pack under body, for good swimming

Webbed feet

PUFFIN

Most auks are colonial breeders, and their nest sites are noisy and filled with activity. The puffin's bright bill becomes even more colorful during the breeding season, to attract and court a mate. Puffins nest in burrows in clifftop soil, including old rabbit warrens. These birds catch small fish such as sand eels, and also shellfish and worms.

Main groups of seabirds include:

Gulls and terns
95 species
• mostly white or gray
• narrow wings

Albatrosses, petrels
93 species
• ocean seabirds
• long, narrow wings

Pelicans, cormorants, gannets
57 species
• gliding and soaring flight
• white or very dark plumage
• feet completely webbed (all four toes)
• throat pouch

Auks
22 species
• compact and dumpy body
• dive well
• mainly black and white plumage

WATERFOWL

 Water birds are built primarily for swimming. They have buoyant bodies and webbed or lobed feet to help push themselves through the water. Many, notably the geese and swans, have long, flexible necks that they can use to reach down into the water as they feed.

SWAN

Swans pair for life, male and female staying together until one dies. They feed in shallow water on weeds, stems, and leaves. Baby swans are called cygnets.

Bar-headed goose

WHITE AND BLACK

The swans are the largest water birds and most have pure white plumage. An exception is the Australian black swan, which is mostly black, with a bright red bill. Ducks tend either to be surface feeders, like the mallard, or divers, such as the pochard and tufted duck. Ducks and geese can reach speeds of up to 55 miles per hour (90 km/h). They also fly high when migrating. The bar-headed goose, for example, has been seen at 30,000 feet (9,000 m) when flying over the Himalayas.

GREBES AND DIVERS

Grebes are more delicate water birds. They build floating nests on lakes and ponds, and have lobed rather than webbed feet.

Divers (or loons) are streamlined water birds and spend most of their time on lakes or at sea. They are very clumsy on land as their legs are set far back on their bodies, perfect for swimming but not for walking!

Small head for size

Ragged crest

HOATZIN

This strange bird spends most of its life clambering in the branches of trees along riverbanks and swamps in South America. It is a poor flier, and although regarded as a water bird, it is more closely related to cuckoos and gamebirds than to swans and ducks. The young hoatzin has a pair of claws on the front of each wing, to help it climb in trees. These are an evolutionary leftover from ancient times. The claws are lost as the hoatzin grows.

Large-clawed feet hold twigs

Main groups of water-birds include:

Ducks, geese, swans
150 species
• most are good swimmers
• webbed feet
• many species are long-necked
• bill usually flattened and quite short

Grebes
20 species
• very good swimmers and divers
• worldwide
• lobed feet

Divers or loons
4 species
• water birds, very good swimmers and divers
• breed in or near the Arctic

93

WADING BIRDS

Birds that wade include herons and storks, and the true waders such as curlews and avocets.

There are about 60 different species of herons found throughout the temperate and tropical parts of the world. A heron feeds by waiting until a frog or fish gets into range, then stabbing suddenly down to catch the prey in its long bill. Long legs help herons, storks, and flamingos feed in quite deep water, without having to swim.

TRUE WADERS

Waders can be found at the water's edge, either at the muddy margins of lakes and ponds, or more typically flocking to the seashore. They feed by probing into soft soil or mud for insects, crustaceans, worms,

*Broad feet
do not sink
in soft mud*

*Spear-shaped bill darts
at fish and frogs*

SANDHILL CRANE
There are 15 kinds of cranes. Most are rare and protected by wildlife laws. They live in flocks except during the breeding season, when they pair off and carry out elaborate, noisy courtship dances.

GREEN-BACKED HERON
This small type of heron waits motionless on an overhanging branch for prey in the water below.

and other small creatures. Plovers, sandpipers, oystercatchers, and curlews are all in this group. Curlews use their long, curved bills to extract worms and larvae from deep below the surface. The bill of the oystercatcher is used like a chisel to pry apart the shells of mollusks such as mussels and oysters.

Broad or spatulate bill

SPOONBILL
The spoonbill's beak widens at the tip, but is not used as a spoon. Instead, it swishes through the water, slightly open, to sieve out small fish and other creatures.

Main groups of wading birds include:

Herons, storks, ibises, flamingos
115 species
• most are freshwater wetland birds
• long legs, long neck, large bill
• most species are large

True waders
200 species
• most are freshwater wetland or seashore birds
• long legs, many have long bill
• medium-sized or small
• some feed in large groups

OWLS

Owls, nightjars, and frogmouths are specialist night hunters. Owls can fly silently and also quite slowly. The soft edging to their feathers helps to cut down the sound of their wings moving through the air. Although they have good vision in the dusk, owls mainly detect their prey on the darkest night by hearing. They can hear four times better than a cat. The smaller owls mostly eat insects and other invertebrates; the larger species take small birds and small mammals.

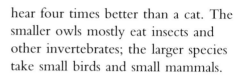

FISHY FOOD

The fishing owls of Africa and Asia feed on fish and amphibians. These owls have long, unfeathered legs and very sharp claws to grab prey as they swoop over the surface.

SNOWY OWL
Most owls have brown, mottled or striped plumage, but the snowy owl is pure white (adult male) or white with dark barring (adult female). This makes the snowy owl hard to see in its snowy Arctic habitat.

NIGHTJARS

Nightjars look like overgrown swallows in shape. But they have patterned brown plumage for excellent camouflage. This makes them almost impossible to spot as they sit among leaves and twigs by day. At night they flit about, snapping up moths in their huge mouths. Some nightjars have long, thin, whiskerlike feathers around their mouths, to feel for prey in the darkness.

Feathers form facial disk that channels sounds into ears

TAWNY OWL
The tawny is one of the most common owls and lives across Europe, North Africa, and northern Asia. Like most owls, it hunts small creatures such as mice, voles, baby rabbits, and small birds. When food is short it also preys on beetles and worms.

Owls
135 species
- most are nocturnal
- soft feathers
- large, flat face and large eyes
- camouflaged plumage

Nightjars and frogmouths
100 species
- nocturnal
- sleek and graceful
- large, dark eyes
- large gape
- camouflaged plumage

Owl records
- The barn owl has the widest distribution of any bird, being found in every continent except Antarctica, and across many different climates and habitats.
- The tiny elf owl from the southern USA and Mexico measures only about 5 inches (12 cm) in height.
- The largest owl is the European eagle owl, which can measure 28 inches (70 cm) tall, with a wingspan of 5 feet (150 cm).

97

BIRDS OF PREY

There are some 295 species of bird of prey (apart from owls), found throughout the world. They range in size from tiny falconets, almost as small as sparrows, to the massive condors, vultures, and sea eagles.

Birds of prey share the features of powerful talons with sharp claws for grasping their prey, a sharp, hooked beak for tearing at flesh, and large eyes giving amazing vision. Many of the smaller birds of prey eat insects, while some, such as the osprey and fish eagles, rely mostly on fish.

THE FASTEST FLIERS

These aerial predators, also called raptors, include some of the world's most acrobatic and speedy fliers. Certain falcons, notably the hobbies and peregrine, are adapted to chase and catch other birds in swift flight. They can twist and turn faster than our eyes can follow.

Long, pointed wings for aerobatic maneuvers

Hooked bill for tearing up victim

PEREGRINE
One of the largest falcons, the peregrine is unequaled in the speed and precision of its flight. It lives in various habitats, usually remote mountains and uplands. The 17 various geographic groups show different plumage, especially in the speckles and bars on the chest.

SPARROWHAWK
As well as sparrows, this woodland predator catches tits and many other small birds. Like many birds of prey, the female is slightly larger than the male.

Bright yellow legs

NOT BALD
One of the most impressive raptors is the bald eagle—which is not bald at all. Its white-feathered head appears naked from a distance. This fish eagle is the national bird of the United States.

MARTIAL EAGLE
The eagles are the largest raptors and take prey up to the size of hares, small deer, and sometimes farm animals such as young lambs.

Birds of prey groups include:

American vultures
7 species
• large
• powerful bill, naked head
• soaring flight

Secretary bird
1 species
• lives on African grasslands
• long legs and crest
• eats reptiles, especially lizards and snakes

Osprey
1 species
• worldwide
• hunts fish

Falcons and relatives
62 species
• worldwide
• small to medium-sized
• long tail

Hawks, eagles, buzzards, vultures and relatives
224 species
• worldwide
• medium-sized to large
• hunt live prey

PHEASANTS AND PIGEONS

Pheasants are long-tailed members of the game bird group, or galliforms. Male pheasants are beautifully colored and patterned, in shades of blue, red, and gold, but the females have drab, camouflaged plumage. This is because the male's main job is to impress and court the female, at breeding time. But afterwards, the female's main task is to raise the chicks while sitting on the nest. So camouflage is very important in these generally woodland or forest birds.

TASTY MEAT

Grouse, turkeys, partridges, quails, and guinea fowl also belong to this group. Many of these are hunted, by other animals as well as people, since they are plump and their meat is very tasty. Doves and pigeons are well known for their soothing calls, and from the tame pigeons which flock in the squares of many towns and cities.

MONAL PHEASANT

Most pheasants and other game birds have stout, heavy bodies and spend much time on the ground searching for plant food. They fly fast and low with rapid, whirring wingbeats, but have little stamina for traveling long distances.

Male monal pheasant in bright breeding plumage

Strong legs and feet for walking and running

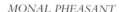

The European cuckoo is the best known of all the cuckoos, but there are more than 125 other kinds around the world. About one third are brood parasites. They lay their eggs in the nest of another species and let the unrelated host do all the hard work of rearing the cuckoo's chick.

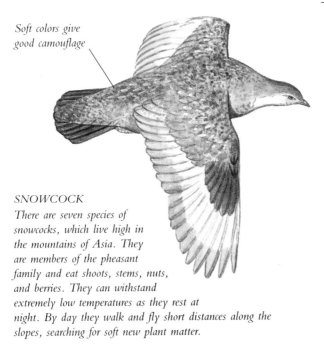

Soft colors give good camouflage

SNOWCOCK

There are seven species of snowcocks, which live high in the mountains of Asia. They are members of the pheasant family and eat shoots, stems, nuts, and berries. They can withstand extremely low temperatures as they rest at night. By day they walk and fly short distances along the slopes, searching for soft new plant matter.

Gamebirds and similar groups include:

Game birds
260 species

Pigeons, doves
300 species

Cuckoos and relatives
150 species

Sandgrouse
16 species

Sandgrouse are grouselike birds of dry desert habitats.

• The males bring drinking water to the chicks by soaking their breast feathers at a water hole and then flying back to the nest.

• The round trip to collect moisture like this from the nearest water hole or oasis may be more than 30 miles (50 km).

PARROTS

Parrots are comical-looking birds with heavy, often hooked bills, and an inquisitive expression and nature. They are intelligent and long-lived, which goes some way toward explaining their popularity as cage birds. Only cage-bred birds should be kept, however, and many parrots die each year during illegal smuggling. Several kinds of parrot are highly endangered in the wild.

CRUSHED NUTS
The bill of a parrot can crush even the heaviest seeds and nuts. But is also used as a kind of third limb, to help the parrot clamber about through the branches of trees.

BUDGIE
The budgerigar is a small species of parrot, native to Australia. Cage-bred budgies come in many colors, but wild budgies are green, and blend in well with the trees and grasses.

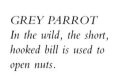

GREY PARROT
In the wild, the short, hooked bill is used to open nuts.

In fact, parrots often prefer to climb rather than fly. They use their strong feet to manipulate their food, as well as for perching and climbing. They also taste and manipulate food items with their long, flexible tongue. Although they mimic sounds in captivity, they are rarely known to do so in the wild.

LIFE IN TREES

Most parrots spend almost all their lives in trees, nesting in holes in tree trunks. They rarely land on the ground or fly long distances except to new feeding grounds.

SCARLET MACAW
This is one of the largest and most brightly colored members of the parrot group, Psittaciformes. However, it has become rare in the wild, due to its capture for the illegal cage-bird trade, and to the destruction of its rain forest homes in Central and South America.

Parrots, cockatoos, and lories

330 species
- chunky birds
- heavy, strong bill
- feet with 2 clawed toes forward and 2 backwards
- mainly green, although some are brightly colored
- mainly tropical or Southern Hemisphere

The cleverest birds?

- Some tame parrots have been taught to identify and ask for (or refuse) more than 50 different objects. They can almost converse with their trainers!
- Parrots are among the longest-lived birds. Ages of 40-50 years are not unusual and some have lived for 80 years or more.
- The flightless kakapo of New Zealand is actually a kind of parrot. So is New Zealand's kea, which lives in upland country 6,600 feet (2,000 m) high, even in snowy weather.

WOODPECKERS

As their name suggests, woodpeckers use their large, sharp, powerful bills to dig into the trunks of trees. They do this to search for insect food such as beetles and grubs under the bark or in the wood, to excavate nesting holes, and to signal to other woodpeckers.

Bill is made of spongy horn and is very light

TROPICAL TOUCANS
The toucans of South and Central America are tropical relatives of woodpeckers, best known for their massive, brightly colored bills.

Kingfishers are also relatives of woodpeckers. They feed on fish, of course, but some of them take insects and reptiles too. Many have bright plumage and darting flight.

Bee-eaters, another related group, are also bright and long-billed. Their agile flight helps them chase and catch flying insects, including bees, wasps, and even hornets, which they then rub on a perch to remove the sting before swallowing.

TOCO TOUCAN
This South American toucan's huge bill is used for feeding on fruit. It also serves as a bright breeding signal, like a highly colored flag, to attract mates in the gloom of the rain forest.

FEMALE GREAT
SPOTTED
WOODPECKER
*The most common
woodpecker across much
of Europe, this species
also ranges through North
Africa and Asia. The male
has a red patch on the back
of its head.*

*Carmine
bee-eater*

HORNBILLS

The hornbills of Africa and
Asia have large bills, like
toucans, and most feed on
fruit. They are bulky birds with
striking black and white plumage.

Although they look very different,
swifts and hummingbirds are also
related. Both groups spend much of
their lives in flight. Hummingbirds sip
nectar by hovering close to flowers.

Woodpeckers and similar bird
groups include:

True woodpeckers
200 species
• range from sparrow-sized to
crow-sized

Toucans
38 species

Barbets
78 species

Honeyguides
15 species

Kingfishers
86 species

Bee-eaters
24 species

Hornbills
45 species

Rollers
16 species

Swifts
74 species

Hummingbirds
315 species

LARGE PERCHING BIRDS

Many birds perch. However, "perching birds" is also the common name for the huge bird group correctly called the passerines. This group contains about three fifths of the 9,000 bird species.

These types are known as perching birds because their feet are supremely adapted to clinging onto a perch, such as a twig or branch. Three toes point forward and one backward. (In many other birds, two toes point forward and two backward).

CROWS
Larger perching birds include members of the crow family, such as crows, jackdaws, magpies, jays, and the very large ravens. Most crows have black or drab plumage, but the jays and magpies are quite colorful. Other larger perching birds are bulbuls, starlings and mynahs, lyrebirds, and orioles.

EUROPEAN JAY
The jay, like other crows, has a selection of raucous, hoarse calls. It is an opportunist, eating many foods, from seeds to beetles.

106

Head crest

BIRDS FROM PARADISE

Perhaps the most spectacular of all the larger perching birds are the birds of paradise. The males are adorned with amazing plumes and colorful feathers. They perform elaborate dances in their Southeast Asian rain forest homes, to attract the females during the breeding season.

WHITE-CHEEKED BULBUL
The many species of bulbul range across Africa and southern Asia.

Adaptable bill can deal with most foods

HILL MYNAH
A noisy and sociable bird, the hill mynah has yellow flaps of skin called wattles on its head. It came originally from southern and Southeast Asia but has been introduced into other regions.

Some of the groups of larger perching birds:

Bulbuls
118 species

Crows, rooks, jays, jackdaws and ravens
116 species

Starlings and mynahs
106 species

Shrikes
70 species

Birds of paradise
43 species

Mockingbirds
30 species

Orioles
28 species

Drongos
20 species

Bowerbirds
18 species

Lyrebirds
2 species

107

SMALL PERCHING BIRDS

Many of the small, familiar birds of gardens, parks, and woodlands around the world belong to the passerine (perching bird) group. They include species such as wrens, tits, finches, sparrows, larks, buntings, and swallows.

This group also includes some of the finest songbirds, such as the song thrush and other thrushes, the famous nightingale, and the many different kinds of warbler.

Buntings are finchlike seed-eating birds, often with brightly colored plumage. In contrast, many warblers have very drab plumage—shades of yellow, green, and brown—which makes them hard to spot. They announce their presence with their loud, pleasant songs and are a major feature of the soundscapes in temperate lands during spring and summer. Indeed, warblers are much more likely to be heard than seen, and some species are identified most easily by their songs—they are so similar in appearance!

THE SMALLEST BIRDS
Hummingbirds are the world's smallest birds. They are similar to many small passerines, but actually belong to the swift group, the apodiforms. Most species have very long, thin bills which they probe deep into flowers, to reach the sweet, energy-packed nectar. Their wings beat so fast that they allow the hummingbird to hover, and they also make the humming noise this bird is named after.

Ruby-throated hummingbird

THRUSHES AND WRENS

Thrushes, which include the American robin, are familiar in the garden. They eat a wide variety of foods, from worms, insects, and other invertebrates, to shoots, seeds, and fruits. Wrens are among the smallest of the group and flit, often unnoticed, in the undergrowth.

WALLCREEPER
Originally a bird of rocky uplands and cliffs, the wallcreeper now searches for insects on the walls of our buildings.

REDSTART
During courtship, the male redstart spreads out his brilliant russet-colored tail feathers. This feature led to the bird's name, "start" from the old word steort, meaning "tail."

Some of the groups of smaller perching birds:

Buntings
552 species

Warblers
339 species

Thrushes
304 species

Flycatchers
156 species

Finches
155 species

Larks, wagtails, and pipits
130 species

Weavers
95 species

Swallows
74 species

Dippers and wrens
64 species

Tits
62 species

Sparrows
35 species

Nuthatches
21 species

Treecreepers
14 species

MAMMALS

When asked to think of "animals," most people think first of mammals. Yet mammals make up only a small number of species in the animal kingdom—about 4,150, compared with 9,000 bird species, 25,000 fish, and more than one million insects.

The key feature of mammals is that the females feed their young on milk produced by the mother. Also, mammals are warm-blooded. (The only other warm-blooded animal group is the birds.) And mammal bodies have fur or hair, to keep in their body warmth.

Mammals that live in the water often have thick layers of fat under their skin, as extra insulation. The mammals also include a group that

110

has truly mastered flight, the bats. (The only other two flying groups are birds and insects.)

BIRTH AND BABIES

Some mammals, the marsupials, give birth to babies that are at a very early stage of development. But most mammals (about nine tenths of species) give birth to their young at a more advanced stage. These are known as placental mammals, after the body part the placenta, which nourishes the growing babies inside the mother's womb.

Some young mammals, such as guinea pigs and antelopes, can run and feed soon after being born. With others, such as rabbits and cats, the young are blind and helpless at first.

LARGE MARSUPIALS

Marsupial mammals are named for the female's marsupium. This is the pocket or pouch on the female's chest or belly, where her young feed on her milk after they have been born. This is why marsupials are sometimes known as pouched mammals.

The best-known marsupials are kangaroos and their smaller relatives the wallabies. These have come to be symbols of their native land,

Australia. They have muscular hind legs and bound along at enormous speed, using their thick tail to balance. The newborn baby kangaroo is only the size of a grape. It crawls through its mother's fur to her pouch, feeds on her milk, and develops rapidly. The older youngster leaves the pouch to explore, but scuttles back if danger threatens.

GRAY KANGAROO
In some areas these kangaroos are pests, eating crops and natural vegetation.

Young kangaroo is called a "joey"

Large tail serves as third leg for resting

112

COMMON IN THE GARDEN

Another familiar marsupial is the opossum of eastern North America. It feigns death—"plays possum"—when it is threatened by attack. It uses its long tail to hang from trees.

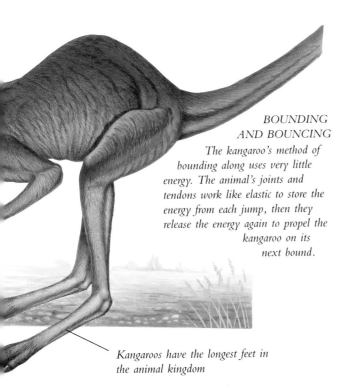

BOUNDING AND BOUNCING

The kangaroo's method of bounding along uses very little energy. The animal's joints and tendons work like elastic to store the energy from each jump, then they release the energy again to propel the kangaroo on its next bound.

Kangaroos have the longest feet in the animal kingdom

Mammals

4,150 species

- young fed on mother's milk
- body with fur or hair
- warm-blooded
- most have four limbs
- breathe through lungs

Main groups of mammals include:

Monotremes

3 species

- the only mammals that lay eggs
- adults are toothless
- Australia and New Guinea
- include platypus and echidnas (spiny anteaters)

Marsupials

270 species

- tiny, undeveloped young raised in mother's pouch
- live mainly in Australia, New Guinea, and South and Central America

113

SMALL MARSUPIALS

The cuddly looking koala is one of the most appealing of all the marsupials. In fact, it has very sharp claws to help it cling onto tree branches and is less cuddly in real life. Koalas spend much of their time snoozing and most of the remaining time munching on eucalyptus leaves. They rarely come down to the ground.

A confirmed ground dweller is the closely-related wombat, which digs a system of underground burrows. Both koalas and wombats look like small bears, but they are not related to true bears, they are marsupials. In fact, marsupials have evolved their own "versions" of many other mammals. There are marsupial moles, marsupial mice, and even marsupial rats.

HONEY POSSUM
This tiny marsupial lives on nectar and pollen from flowers, collected with its long snout and brush-tipped tongue.

114

KOALA AND YOUNG
Koalas are slow-moving and inoffensive—unless threatened, when they slash out with their sharp claws. They eat almost nothing but eucalyptus (gum) leaves. They hardly ever drink, obtaining moisture from the leaves.

GLIDERS
Several marsupials, known as gliders, can swoop in the air from tree to tree. They do this by using a parachute-like flap of skin stretching from the front to hind leg on either side of the body. The glider spreads out these flaps as it jumps.

MARSUPIAL MOLE
The female's pouch opens backward, so it does not fill with soil as she tunnels.

Some of the main marsupial groups:

American opossums
75 species

Marsupial mice and relatives
53 species

Kangaroos and wallabies
50 species

Possums
23 species

Bandicoots
17 species

Brushtail possums and cuscuses
14 species

Ringtail possums, pygmy possums and gliders
30 species

Rat kangaroos
10 species

Wombats
3 species

Koala
1 species

115

SMALL RODENTS

The largest group by far of the placental mammals is the rodents, with some 1,600 species. They include hundreds of different species of mice, rats, voles, lemmings, jerboas, squirrels, chipmunks, beavers, and many others. A rodent's key feature is its sharp front teeth (incisors) that gnaw and chew through even the hardest foods, such as nuts or tree bark.

Some small rodents, for example mice, rats, hamsters, gerbils, and guinea pigs, have been domesticated and are often kept as pets.

MIGRATING LEMMINGS
Lemmings live in the cold northern lands of the tundra. Sometimes their populations build up to such a level that the lemmings eat all the food. So they must move on or migrate, often in huge numbers.

DORMOUSE
Some rodents hibernate, sleeping deeply through the cold season. The dormouse makes a grassy nest among tree roots. Its body becomes so cold that its ears and feet turn purple.

116

SMALL TO BIG

There are about 775 different kinds of mice and rats, ranging from the tiny pygmy mouse, which has a total length of 4 inches (10 cm), to large rats eight times as long. The harvest mouse is very agile and uses its long tail as an extra limb when clambering among grass stems.

NORWAY LEMMING
Lemmings stay active through the winter, digging tunnels under the snow to reach berries and seeds.

Jerboa

Dormouse (out of hibernation)

Rodents

1,600 species

- small to medium-sized placental mammals
- strong, chisel-shaped incisor teeth at front of mouth, for gnawing
- many are active at night
- very widespread and numerous (the house mouse is the most widespread mammal after the human being)

Main groups of smaller rodents include:

Pocket mice

65 species

Rats, mice, voles, hamsters, gerbils, and relatives

1,082 species

Dormice

10 species

Jerboas, jumping mice, and birch mice

45 species

117

LARGE RODENTS

Most rodents are small or medium-sized, but several groups contain larger species. Squirrels are adept climbers and spend most of their lives up in the trees, clinging on with sharp claws and using their furry tails to balance. "Flying" squirrels cannot really fly, but glide well from tree to tree, using the stretched skin between front and back limbs.

Marmots and prairie dogs are like chunky, tail-less squirrels. They live in underground burrows. Beavers can measure 4 feet (120 cm) in total length. They swim well, using their webbed hind feet and flat rudderlike tail.

BEAVER
Beavers gnaw at trees to feed on the soft bark and young sap-rich wood just beneath. They cut down trees to eat the higher branches and also to use the logs for dam-building.

PORCUPINES

The porcupines have sharp spines, which are very thick hairs used for defense against predators. American porcupines live in trees and have long, prehensile tails to help them cling onto branches. European, African, and Asian porcupines are ground dwellers.

LODGE
The beaver's home is called a lodge. It is built in a lake created by the beavers damming a stream.

RED SQUIRREL
This type of squirrel has become rare in parts of Europe, partly because its natural conifer woodland home has been replaced by conifer plantations.

Groups of larger rodents include:

Squirrels, marmots, and relatives
267 species

Porcupines
21 species

Cavies (guinea pigs)
14 species

Agoutis
13 species

Chinchillas
6 species

Beavers
2 species

Capybara
1 species

Coypu
1 species

Springhare
1 species

Largest rodent
• The capybara of South America measures up to 55 inches (135 cm) long and 24 inches (60 cm) tall at the shoulder.

ELEPHANTS

There are only two species of elephant: the African and the Asian. The African elephant has a larger body, bigger ears, longer tusks, and a less bumpy forehead.

African elephants live south of the Sahara Desert in savanna, forest, or dry scrubland. They roam in small family groups feeding on grass, bark, and twigs. They move slowly and steadily most of the time, walking at about the same speed as us. But a charging elephant can outrun a person, reaching a speed of 28 miles per hour (40 km/h) for a short sprint. The rumbling growl of an elephant can carry for more than a mile through the bush, and elephants use their deep voices to keep in touch. The tusks of an elephant are oversized incisor teeth and get larger through life.

Trained Asian elephant

Trunk is elongated nose and upper lip

Tusks trimmed for safety

AFRICAN ELEPHANT
Large male elephants usually live alone except during the mating season. If they feel in danger, they face the enemy and extend their ears to look even bigger than usual!

ELEPHANTS IN DANGER

Many elephants are illegally hunted and killed so that their tusks can be turned into the white substance ivory.

Elephants use their trunks for many different jobs—smelling friends, sniffing for enemies, plucking leaves from trees, and greeting other elephants by strokes and caresses.

Elephants

2 species

- African elephant is found scattered in Africa, mainly Central, East, and South
- Indian elephant is found in India, Sri Lanka, South China, Southeast Asia

Elephant records

- Elephants are the largest land animals. A big male African elephant can grow to almost 13 feet (4 m) tall at the shoulder and weigh more than 5 tons. His tusks can be over 10 feet (3 m) long.
- Elephants cool themselves down in hot weather by gently flapping their ears. The heat is lost to the air over a huge surface area from the blood vessels in the ears, which work like heating radiators in reverse.

121

BIGGER CATS

Lions are probably the best known of all the big cats, but there are six other members of the true big cat group: the tiger, jaguar, leopard, snow leopard, clouded leopard, and cheetah.

Most of the big cats are either spotted or striped, but the adult lion is a uniform sandy color, although lion cubs are spotted. The black panther is actually a very dark color variety of the leopard. Black jaguars also occur. Big cats are active hunters of live prey, usually other mammals or birds.

LIVING TOGETHER

Lions are the only cats that live together, in family groups known as prides. They can tackle animals larger than themselves by working together to ambush their prey. They join in and gorge themselves on the meat, before sleeping it off until the next hunting foray.

BIG AND BIGGER
Four of the seven species of true big cats are shown here. So are some other cat species— including the mountain lion (also known as the puma or cougar) and the lynx.

Black panther (a black variety of leopard)

Leopard

Lynx

Cheetah

The jaguar is unusual among cats, in that it eats a very varied diet that includes deer, monkeys, sloths, rodents, birds, turtles, frogs, and plenty of fish.

The thick fur of the snow leopard gives it excellent insulation in the icy mountains of Asia, at well above the timberland.

Mountain lion

Jaguar

Lions

Big cats
7 species
- Lion, lives in Africa, mainly in grasslands; very small population in India
- Tiger, lives in India, China, Indonesia, in forest, swamp, and scrub
- Jaguar, lives in South and Central America, mainly in damp forests and swamps
- Leopard, lives in Africa and Southern Asia, in a wide variety of habitats from mountains to forests, swamps, and semidesert
- Snow leopard, lives in Asia, in uplands and mountains
- Clouded leopard, lives in Southeast Asia, in forests
- Cheetah, lives in Africa and south west Asia, mainly in grassland and dry scrub

Biggest big cat
- The Siberian type of tiger measures more than 10 feet (3 m) from head to tail, and weighs over 650 pounds (300 kg).

SMALLER CATS

Whatever their size, big or small, most cats show similar behavior, especially when hunting. They work at night and alone. The cat stalks its prey quietly, keeping low near the ground. It then makes a quick dash to grab the victim, holding it down with sharp claws, biting and slashing. The cat often clamps its jaws and teeth onto the prey's neck, to close its windpipe until it suffocates.

CAT CLAWS

Cats are unusual among mammals in being able to pull their claws into fleshy toe sheaths when the claws are not needed. This keeps the claw points clean and sharp —and makes pet cats more comfortable to handle than dogs!

MARGAY
One of about 20 species of smaller cat, the margay lives in Central and South American forests and scrub, hunting small animals such as rabbits, squirrels, rats, and birds.

DOMESTIC CATS
The Abyssinian breed of domestic or pet cat is probably very similar to the original wild cats, from which all pet cats have been bred. The first domestic cats may have scavenged around human settlements more than 9,000 years ago.

124

BIGGER SMALL CATS

The lynx (shown on page 122) is one of the biggest small cats. It has well padded, heavily furred feet and long legs—adaptations that help it move through deep snow. The largest of the "small" cats is the puma (also shown on

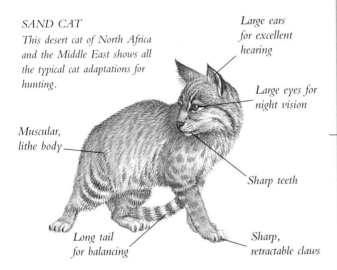

SAND CAT
This desert cat of North Africa and the Middle East shows all the typical cat adaptations for hunting.

Large ears for excellent hearing

Large eyes for night vision

Muscular, lithe body

Sharp teeth

Long tail for balancing

Sharp, retractable claws

page 122). It may be more than 6 feet (1.8 m) long including its tail. Large pumas can even kill fully grown deer. Most species of smaller cats live in forests and have spotted coats for camouflage—which has led to themselves being hunted by people for their furs.

Smaller cats

28 species, including:

- mountain lion (puma or cougar) from North and South America
- lynx from Europe, Asia, North America
- bobcat (red lynx) from North and Central America
- ocelot from North and South America
- serval from Africa
- wild cat from Europe, Africa, and India

More cat records

- The cheetah is the fastest cat, and the fastest of all mammals over short distances—it can sprint at almost 60 miles an hour (100 km/h), but not for long.
- The smallest wild cat (as opposed to domesticated cats) is probably the rusty-spotted cat of India, which is about 28 inches (70 cm) long, including the tail.

125

WOLVES AND FOXES

The dog family includes wolves, foxes, and jackals. The familiar domesticated dogs, which have been bred into a huge number of different shapes and sizes, probably originated from the gray or timber wolf. Unlike cats, dogs tend to hunt in groups, called packs. Their long, powerful legs enable them to run well and for long periods.

The gray wolf is the largest of the dogs, found in forest, tundra and mountains. Wolves eat a wide range of food, from deer to small mammals, and even berries and other fruits. Even though they are common in some places, wolves are very rarely a danger to people.

RED OR COMMON FOX
This is one of the most adaptable and widespread of all mammals, living in remote country and also scavenging in towns and cities.

Australian wild dog or dingo

CALL OF THE WILD

Coyotes are common in parts of North America. They are rather like smaller versions of wolves and their eerie night howls are the "call of the wild." The maned wolf lives in tall grasslands in South America. It has long, graceful legs and attractive red-brown fur.

DOGGY MOODS
Like pet dogs, wolves
put their ears back
and bare their
teeth when
they feel in
danger.

Dogs and relatives

35 species, including:

- Gray wolf, lives in North America, Europe, and across Asia
- Red wolf, from south-eastern North America, now exceptionally rare
- Maned wolf, from South America
- Coyote, in North and Central America
- Jackals, 4 species in Africa, southeast Europe, and southern Asia
- Dingo, from Australia
- Dhole or Asian wild dog, in India, China, and Southeast Asia
- African wild dog, mainly in southern Africa
- Foxes, 21 species, worldwide except Australia

Wolves of the world

- The common or gray wolf varies in appearance across its range and is also called the timber wolf, steppe wolf, tundra wolf, and plains wolf.

BEARS

Bears are large, powerful mammals that feed on a very wide range of foods, from small mammals and fish, to fruits, seeds, and roots. Most bears live in woods and forests. Different species are found from the Arctic right through to the tropical forests, mainly in the Northern Hemisphere.

The biggest bears are the northern types of the brown or grizzly bear, and the polar bear. These huge animals are sometimes a danger to people, especially when protecting their young or cubs. European brown bears are smaller and found in a few isolated mountains.

CREAMY CAMOUFLAGE
The creamy white fur of the polar bear blends in well with its snowy and icy habitat, and undoubtedly helps this species to stalk seals on the ice

BEAR OF MANY NAMES
The brown bear is known by different names in different regions, including the grizzly, Eurasian brown bear, Kodiak bear, and Alaskan bear.

SNOW DEN
The mother polar bear rears her cubs in a cave she digs in the snow. She may not leave it for weeks, surviving on her supplies of body fat.

Asian or Himalayan black bear

without being spotted too soon. As well as thick fur, polar bears have a thick layer of fat under their skin, to keep out the cold. The spectacled bear is the only bear found in South America and takes its name from the white framelike markings around its eyes.

Bears
7 species:
- Polar bear, on the Arctic tundra, coasts, and ice floes
- Brown bear, across North America, Europe and Asia, in woods and forest
- American black bear, from North American woodland
- Asian or Himalayan black bear, in Southeast Asian scrub and forest
- Sun bear, from the forests of Southeast Asia (especially Malaya)
- Sloth bear, in India and Sri Lanka, lives in forests
- Spectacled bear, in South America, especially the damp forests in the Andes

Bear records
- Smallest is the sun bear, about 4 feet (1.2 m) long.
- Largest is the polar bear, at 10 feet (3 m) long and weighing more than 1,300 pounds (590 kg).

SMALLER CARNIVORES

The mongooses and their relatives, the civets and genets, are slender-bodied carnivores found in Europe, Africa, and south Asia. In build they seem to be a combination of weasel and cat. The common genet has a spotted body and striped tail. It comes out at night and climbs trees as it hunts for birds, reptiles, and other prey. Genets are found as far north as France, but are rarely seen.

TOO FAST

Mongooses live mainly in dry country and include snakes in their diet. They even eat poisonous snakes such as cobras. The mongoose is not especially immune to the poison. It uses its lightning reflexes and agility to avoid the snake's fangs.

One of the most famous kinds of mongoose is the

MEERKATS
Also called suricates, these mongooses live in very dry areas, even the Kalahari and Namib Deserts.

130

gray meerkat of South West Africa. Meerkats live in colonies and post individuals as lookouts. These guards suddenly sit bolt upright when they sense danger, and alert the other members of the group.

VEGETARIAN CARNIVORES?
The red panda is related to the giant panda, but is much smaller, with deep red fur and a foxlike shape. In common with its more famous relative, the giant panda, it lives in China, and

RACCOON
Raccoons are well-known in North America, where they raid dustbins for scraps, mostly at night. They use their front paws rather like hands to find and manipulate food.

also in some other Himalayan countries. It, too, prefers bamboo forests. Although these pandas are vegetarian in diet, they are included in the carnivore mammal group because of their body similarities with other carnivores.

Some groups of smaller mammal carnivores:

Mongooses, civets, and relatives
66 species
Including:
True civets and genets
9 species
Mongooses
27 species

Raccoons, pandas, and relatives
17 species
Including:
Raccoons
6 species
Coatis
3 species
Kinkajou
1 species
Pandas
2 species

MUSTELIDS

The mustelids are a large group of smaller mammalian hunters with long bodies, fast reactions, and voracious appetites for meat. The weasel is a specialist hunter of small rodents such as voles and mice, though it occasionally eats birds, frogs, and even young rabbits. Its small size, slim body and narrow head enable the weasel to follow prey as small as mice into their burrows. Stoats are about twice the size of weasels and have a black tip to their tail.

Sable

MORE MUSTELIDS

Pine martens are exceptionally good at climbing and can even catch squirrels. The related sable is found in Siberian forests. It resembles the pine marten but has longer legs and larger ears. The sable produces a fine quality fur and has been both hunted and bred on farms for its pelt.

BADGER

Each badger's face stripes are slightly different, allowing these animals to recognize each other even in darkness. Badgers dig large burrows called setts (or sets), with different chambers for sleeping, resting, and eating. They keep the sett very clean, changing the bedding of grass, leaves, and moss every few days.

THE GLUTTON

Wolverines are the largest mustelids. They have broad footpads and move quickly even over loose snow. Wolverines have a legendary appetite, earning them the nickname "glutton," and kill prey as large as a reindeer. They also eat carrion and cover up to 25 miles (40 km) each day in search of food.

Badgers are also large mustelids. They are mainly nocturnal. They live in family groups but usually forage individually, following well-used tracks. Their food includes earthworms, beetles, small mammals, carrion, roots, and fruits.

WEASEL
This tiny yet fierce carnivore is so slim, it is supposed to be able to slip through a ring that would fit snugly on a human finger.

White underside

More groups of smaller mammal carnivores:

Weasels, stoats, polecats, ferrets, martens, mink, and similar predators (mustelids)
33 species

Skunks
13 species

Otters
12 species

Badgers
8 species

Another group of related carnivores:

Hyenas
4 species
- spotted hyena lives in Africa south of the Sahara desert
- striped hyena lives in Africa, Middle East, and southern Asia to India
- brown hyena lives in Central and Southern Africa

133

BATS

The bats, or chiropterans, form the second-largest mammal group in terms of numbers of species, after rodents. They are the only mammals that can truly fly. There are two main types of bats. These are the large, often day-flying fruit bats and flying foxes, and the mostly small, insect-eating bats that are nearly all nocturnal (active at night).

BAT FOOD

A few bats have specialized foods. Some catch small birds, others swoop over water to grab fish. Vampire bats feed on blood, but, contrary to legend, they are rarely dangerous to people. Vampire bats live in Central and South

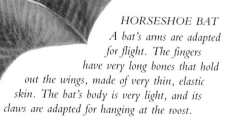

HORSESHOE BAT
A bat's arms are adapted for flight. The fingers have very long bones that hold out the wings, made of very thin, elastic skin. The bat's body is very light, and its claws are adapted for hanging at the roost.

134

ECHOLOCATION

Most bats find their way in the dark by echolocation. They send out very high-pitched (ultrasonic) clicks and squeaks. These reflect or bounce off objects. The bat hears the returning echoes and works out the distance, size, and shape of the objects.

America, are quite small, and only one kind attacks mammals such as sleeping cattle, pigs, or horses. The bat makes a small wound with its sharp teeth, then licks up the blood that oozes out. Most bats rest and sleep in colonies, hanging upside down.

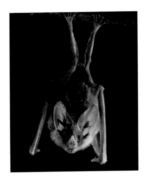

Roosting bat

Fox-eared bat *Horseshoe bat* *Vampire bat*

BAT FACES
Each main type of bat has slightly different facial features, linked to the way it feeds or finds its way by echolocation.

Bats
960 species
- only mammals with true powered flight
- front limbs are modified as wings
- use echolocation in flight
- most are nocturnal

Some groups of bats:

Flying foxes and fruit bats
175 species

Common or vesper bats
320 species

Spear-nosed bats
140 species

Free-tailed bats
90 species

Horseshoe bats
70 species

Leaf-nosed bats
60 species

Vampire bats
3 species

135

INSECT EATERS

The insect eaters or insectivores are mostly rather small mammals, and include shrews, moles, and hedgehogs. Most are nocturnal, and their sensitive noses help them find their way about and track down prey.

SHARP TEETH

The teeth of shrews and similar insectivores are quite different from those of mice. Whereas rodents have long, chisel-like, cutting teeth, a shrew has very sharp teeth, more like those of a miniature cat. Shrews are so small, yet so active, that they must eat food every few hours, or they starve to death.

Moles are well adapted to their subterranean life, with a cylindrical body, short limbs, and soft, silky fur that allows them to slither easily through the soil. Moles use their spadelike front limbs to dig their underground tunnels. Every so often,

COMMON HEDGEHOG
The spines of a hedgehog are modified hairs, and a fully grown hedgehog has about 16,000. They normally lie flat along the body, but they can be raised, using powerful muscles, when the hedgehog is alarmed. With its head and legs tucked into its belly, the hedgehog rolls into a spiny ball. This deters most predators but is, sadly, no defense against road traffic!

they heave the loose soil to the surface through vertical tunnels to make the familiar molehills. A mole's tunnel system may stretch for up to 500 feet (160 m). A much larger molehill, the fortress, covers the mole's resting and breeding chamber.

WATER SHREW
An excellent swimmer, the water shrew hunts tadpoles, small fish, and young frogs, as well as worms.

MOLE
The mole's tunnel system is its feeding place. The mole patrols the tunnels several times daily, eating worms, beetles, and other insects that have emerged through the walls. It stores excess food in a special larder chamber.

Insect eaters (insectivores)
345 species
- eat mainly insects, but also other small animals such as worms, spiders, and slugs
- mostly small and active
- long, narrow snout
- small limbs and ears
- many are nocturnal, with large eyes and long whiskers

Groups include:
Shrews
246 species
Tenrecs and otter shrews
33 species
Moles and relatives
29 species
Golden moles
18 species
Hedgehogs and moonrats
17 species
Solenodons
2 species

ARMADILLOS AND SLOTHS

These animals are often known as the toothless mammals: they either lack teeth altogether, or their teeth are small and simple, and not firmly rooted. Anteaters have a long snout, long sticky tongue and sharp claws. They rip open the nests of ants or termites, then lick up the insects in their hundreds. Armadillos resemble anteaters that have been armor-plated.

Their skin is covered by protective bony plates, and they can tuck in their limbs and roll up into a hard ball if attacked.

A SLOW LIFE

Sloths take their name from their very slow movements. These inhabitants of South American rain forests spend most of their lives suspended upside-down from tree branches, feeding on

ARMADILLO
This armored mammal eats ants, termites, and other small invertebrates, and also other food such as fruit. It lives in South and Central America, and the southern US. The smaller pink fairy armadillo, only 5 inches (12 cm) long, dwells in South America.

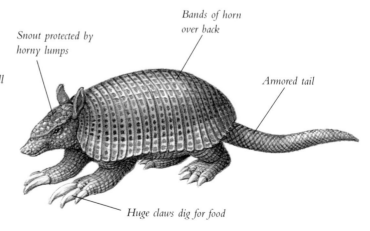

Snout protected by horny lumps

Bands of horn over back

Armored tail

Huge claws dig for food

Three-toed sloth

fruit and leaves. They are often disguised among the leaves by the tiny green plants (algae) growing in their dense fur. Even the sloth's digestion is slow. It can take a month for a meal to pass through the sloth's body.

Hooked claws cling to branches

PANGOLIN
This nocturnal climber searches for ants and termites among the branches. Pangolins live in Africa, southern and Southeast Asia.

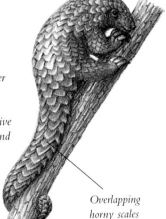

Overlapping horny scales

Toothless mammals
29 species
• slow-moving
• most have long claws
• very simple teeth, or no teeth

Groups include:
Armadillos
20 species
Sloths
5 species
Anteaters
4 species
Pangolins
7 species
Aardvark
1 species

Long tongue
The extremely long tongue of the giant anteater can collect hundreds of termites or ants with a single lick. The anteater may eat tens of thousands of ants in a single day.

RABBITS AND HARES

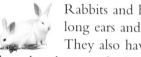
Rabbits and hares have long ears and short tails. They also have long back legs that they use for hopping and also for running and leaping at great speed. Although they resemble rodents, rabbits and hares have their own mammal group, lagomorphs, meaning "leaping shapes."

LIFE IN THE OPEN
Many rabbits and hares live in open, grassy country where they need their speed to take them clear of their enemies. Hares are the fastest, reaching speeds of 50 miles per hour (80 km/h). They have strong teeth and can gnaw and chew hard seeds, roots, and bark. The position of the eyes, on the sides of the head, gives rabbits and hares good all-round vision, even when keeping their heads perfectly still. They can even see

EUROPEAN RABBIT
This rabbit has become a serious pest in many parts of the world (see page 240). In one year, a single female can produce more than 20 young, and these begin to breed themselves when only four months old. However, many animals hunt rabbits, from eagles to foxes and stoats.

behind themselves. Their large ears can quickly pick out any sound and they are ever alert to danger. Rabbits tend to live in systems of tunnels dug into the soil, but most hares have their babies on the surface of the ground.

Pikas are small, rabbitlike mammals with rounded ears and almost no tail. They live in mountainous country in Asia and northwest North America.

COTTONTAIL
This rabbit rarely digs its own burrow. It takes over another animal's, or rests in a sheltered place on the surface, called a form.

Rabbits and hares (lagomorphs)
44 species
- large ears
- long hind legs
- eyes face sideways
- very short tail
- slit-like nostrils

Pikas
14 species
- rounded ears
- almost tail-less
- resemble large voles

Rabbit or hare?
- The American jackrabbit is actually a species of hare.
- Like other hares it has large ears. It uses these to hear, and also to regulate its temperature. In hot weather the ears give off body warmth, keeping the hare cool.
- In North America, some kinds of rabbits are known as cottontails.

SEA MAMMALS

Whales and dolphins are the mammals most perfectly adapted to life at sea. They never come ashore (unless beached by accident) and cannot survive on land. The baleen whales, which are mostly large, feed by filtering tiny organisms from the sea. The toothed whales include the sperm whale, which eats mainly squid, and the many kinds of dolphins and porpoises.

SEAL OR SEA LION?

Seals and sea lions are also at home in the open sea, but they come ashore to bask and breed, either on rocky coasts or on open ice. Seals eat a range of prey, from fish and birds to crabs and squid.

The true seals have back flippers that cannot be bent under the body to walk on land. They slide or slither on the ground, but swim with

SEA LION
Most species of sea lions live in groups. They come ashore at traditional breeding places called rookeries, where the males battle with each other to win the attention of females.

supreme grace by bending their bodies from side to side. Sea lions use their long front flippers more when swimming, and they can shuffle about on their leglike hind flippers when on land.

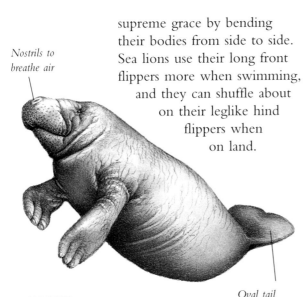

Nostrils to breathe air

Oval tail fluke

MANATEE
The manatees and dugongs (both known as seacows) are gentle herbivores. They swim slowly in tropical seas and estuaries, grazing on water plants and algae.

TUSKED SWIMMER

The walrus is one of the giants of the group. It has a very fat body that keeps out the cold of the Arctic Ocean. A large male walrus may be almost 13 feet (4 m) long. It uses its long tusks to lever shellfish from the seabed.

main groups of marine (sea-dwelling) mammals:

Whales and dolphins (cetaceans)
76 species
Includes:

Toothed whales, dolphins, and porpoises
66 species

Baleen or great whales
10 species

Seals and sea lions (pinnipeds)
34 species
Includes:

True seals
19 species

Fur seals and sea lions
14 species

Walrus
1 species

Manatees and dugong (sea cows or sirenians)
4 species

PIGS, RHINOS, AND HIPPOS

The hoofed mammals, or ungulates, range in size from tiny deer to huge rhinos and hippos. Their key features are toes that end in clublike hooves rather than sharp claws. They are divided into two main groups, the even-toed or cloven-hoofed ungulates (artiodactyls) and the odd-toed ungulates (perissodactyls).

GRUBBING ABOUT

Pigs are well-known farmyard animals, but the wild ancestors of the domestic pig are the wild boars—which look rather different! Wild boars have much thicker coats and larger heads. They are still common in forests in many parts of Europe and Asia.

Most wild pigs have tusks, formed from up-curved canine teeth. They grub about in the soil for roots, bulbs and small animals. Peccaries resemble wild pigs and are found in forests of South and Central America.

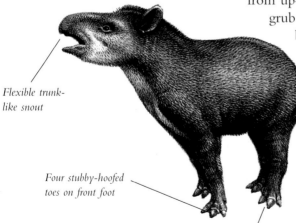

Flexible trunk-like snout

Four stubby-hoofed toes on front foot

Three hoofed toes on back foot

TAPIR
Three of the four tapir species live in South America. The fourth comes from Southeast Asia. Tapirs are active at night, eating leaves, fruits, and other plant food in their forest homes.

RHINOS
The horn of a rhino is not bone or even horn. It is a dense matted mass of compressed body hairs. Most species of rhinos, especially the Asian ones, are extremely rare.

HIPPOS, BIG AND SMALL

There are two different kinds of hippos, both living wild in Africa. The larger species, familiar in wildlife parks and zoos, spends much of its life submerged in rivers or lakes. It emerges at night to graze on soft waterside plants and grasses. The other, the much smaller pygmy hippo, is a rare animal of swampy forests.

WILD BOAR
Stocky and powerful, the wild boar may charge if provoked. Its slashing tusks can cause deep wounds.

Hoofed mammals (ungulates) are divided into:

Odd-toed (perissodactyls)
These consist of:
Rhinos
5 species
Tapirs
4 species
Asses, horses, and zebras
7 species

Even-toed (artiodactyls)
These consist of:
Camels and llamas
6 species
Wild pigs
9 species
Peccaries
3 species
Hippos
2 species
Giraffes
2 species
Deer
41 species
Wild cattle, antelopes, sheep, and goats
124 species

145

DEER AND ANTELOPES

Deer and antelopes are long-legged, browsing or grazing, hoofed mammals. They have the typical flattened teeth of a herbivore, for grinding up plants, and special stomachs to help them digest their food. They tend to live in herds for safety and they can run fast when chased. The main feature of deer is their antlers,

Broad-bladed antlers

which can be many-branched and are shed and regrown each year. In most species it is only the male deer which carries antlers, but in reindeer both sexes have them.

NOT ANTLERS
Antelopes have horns, not antlers, and these grow throughout their lives. In some antelopes the horns are very short, but in others, such as the sable antelope, they are long and curved.

MALE FALLOW DEER
These deer feed mainly at dawn and dusk. During the middle of the day and night, they rest in undergrowth or among trees, where their spotted coats provide camouflage. The background coat color varies from very dark brown, through chestnut and pale fawn, to nearly white.

Antelopes are well known for their speed and agility, and their keen eyesight usually gives them early warning of an attack by their predators. Some antelopes confuse their pursuers by rapid changes of direction, or by suddenly jumping vertically upward.

Reindeer with newly grown antlers

Ringed horns

IBEX
A goatlike animal, the ibex lives at heights of up to 10,000 feet (3,000 m) in the Alps of Europe.

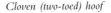
Cloven (two-toed) hoof

Some groups of even-toed ungulates:

- **Deer**
  41 species
  • Includes red deer, fallow deer, sika, roe, reindeer or caribou, elk or moose, muntjac

- **Gazelles and dwarf antelopes**
  30 species
  • Includes Grant's gazelle, Thomson's gazelle, dikdik

- **Grazing antelopes**
  24 species
  • Includes pronghorn, waterbuck, wildebeest or gnu, impala, sable antelope, oryx

- **Spiral-horned antelopes**
  9 species
  • Includes kudus and elands

- **Pronghorn**
  1 species

147

WILD CATTLE

The mighty bison once roamed the forests over the whole of Europe, but retreated to remote wooded areas as the land increasingly came under cultivation. By the beginning of this century they were almost extinct, but have now been reintroduced in some places.

The closely related North American bison or buffalo was similarly almost exterminated in its prairie home, before being rescued.

WILD CATTLE
Until the seventeenth century, there were also wild cattle in Europe and Asia. They were a species known as the auroch—the ancestor of all domesticated cattle.

Cattle and goats probably began their domestication about 9,000 years ago, for their milk, meat, fur, and hides. This means they are among the earliest of all domesticated animals.

TAKIN
Also known as the golden-fleeced cow, this small species of wild cattle stands only 3 feet (1 m) tall at the shoulder. It lives in the high, dense, remote bamboo forests of Southeast Asia and China.

Small horns

Thick, shaggy, golden fur

Powerful legs

LONGEST FUR

Musk oxen have extremely thick, furry coats to protect themselves from the fiercest Arctic storms. Their hair is longer than that of any other wild animal, reaching 36 inches (90 cm) on the neck and flanks.

YAK
The bulky yak lives in the Tibetan highlands, at heights up to 20,000 feet (6,000 m). Herds of yak have been domesticated by local people.

Spiral horns

Himalayan markhor

More groups of even-toed ungulates:

Cattle and relatives
23 species
• Includes domestic cattle, yak, water buffalo, African buffalo, American bison (buffalo), European bison

Sheep, goats and relatives
26 species
• Includes saiga, musk ox, chamois, ibex, Spanish ibex, markhor, mountain goat, wild goat, Barbary sheep, mouflon, bighorn sheep

Biggest horns
The water buffalo, native to India and Southeast Asia, has the largest horns of any living animal. The record spread, from one horn-tip to the other, is more than 13 feet (4 m).

GIRAFFES AND HORSES

Two further groups of hoofed mammals or ungulates are the horses, asses, and zebras (odd-toed), and the giraffe and closely related but very rare okapi (even-toed).

Domesticated horses are familiar on farms and ranches, racetracks, and showgrounds. All the breeds, from tiny Shetlands to giant Shires, belong to one species, *Equus caballus*. They are probably descended from wild horses similar to today's only semiwild species, Przewalski's horse of Mongolia. Zebras are very similar to horses in build and habits, being wary, long-legged, fast-running animals of open grassland.

THE TALLEST MAMMALS

The giraffe and the okapi are both very unusual animals. Giraffes are easy to spot in their open savanna habitat

WHY STRIPES?
Each of the three kinds of zebra—plains, mountain and Grevy's—has a different stripe pattern. The stripes are like fingerprints, unique in each animal. They may help individuals in the herd to recognize each other, and with camouflage in the long African grass.

in Africa, as they tower above other animals on their long legs, using their long necks and long tongues to reach the topmost twigs and leaves.

The related okapi, also with a longish neck, looks like a combination of giraffe and zebra. It is a shy inhabitant of West African rain forest and is seldom seen in the wild.

GIRAFFE VARIETIES
There are several different varieties of the single giraffe species, each with a different brown and white coat pattern.

More groups of hoofed mammals (ungulates):

Horses, asses, and zebras
7 species
Including:

Horses
2 species
• Domestic
• Przewalski's

Wild asses
2 species
• African
• Asian

Zebras
3 species
• Plains
• Mountain
• Grevy's

Giraffes
2 species
• Giraffe
• Okapi

Record height
• Giraffes may grow to almost 20 feet (6 m), making them the tallest mammals.

CAMELS AND LLAMAS

 The camel group of even-toed hoofed mammals includes two kinds of camels. It also includes the domesticated llamas and alpacas, and the wild guanacos and vicunas, which all live in South America.

Wild two-humped or bactrian camels can still be found in some parts of the remote steppe grasslands of Mongolia. They live in dry, sandy habitats and can go for weeks between drinks.

The one-humped or dromedary camels of southwest Asia and North Africa are mostly domesticated, used for centuries for transportation, milk, and meat.

A camel's hump stores not water, but body fat, to help the animal through times of food shortage.

Domesticated llama

PRECIOUS WATER
Camels do not have a magical ability to store water. But they can survive long periods between drinks. When they do find water, they can take in 13 gallons (50 l) at once, since the camel's stomach can expand enormously. They also lose very little moisture as sweat, and produce concentrated urine, which also saves water.

WOOL AND TRANSPORTATION

In South America, both the llama and alpaca have been domesticated as pack animals and also for their wool. The vicuna is a smaller version of the llama and lives at heights of more than 13,000 feet (4,000 m) on the Andes Mountains. It was rare but is now a protected species.

DESERT LIFE
The camel has many adaptations to desert survival. Most one-humped or dromedary camels are domesticated or semiwild.

Large eyelashes keep out windblown sand

Food store in hump

Wide feet do not sink in soft sand

Nostrils can be almost closed in sandstorms

Thick fur protects against sun's heat

Camels and llamas

6 species

- Dromedary, Arabian or one-humped camel, is mainly domesticated, lives across North Africa and the Middle East; also populations introduced as pack animals in other dry places, such as Australia; now run semiwild

- Bactrian or two-humped camel, lives semiwild or domesticated on Mongolian grasslands

- Llama lives in the Andes region of South America, in grassland and shrubland; used for transportation, meat, wool, milk, and skins

- Alpaca is similar to llama but smaller in size; lives slightly higher in the Andes

- Guanaco is widespread at lower altitudes through South America

- Vicuna lives on high Andes, in Peru, Bolivia, Chile, and Argentina

153

MONKEYS AND LEMURS

The lemurs, monkeys and apes make up the mammal group called primates. They have a large brain, well-developed hands with a flexible thumb for grasping objects, and large eyes that both face forward, giving good vision to judge distances accurately.

OLD WORLD MONKEYS

More than 80 different kinds of monkey live in Africa and Asia. One of the most common in Africa is the vervet monkey, which lives equally on the ground or in trees.

Baboons are large, ground-dwelling monkeys with doglike faces and sharp teeth. Although most monkeys live in tropical areas, some, such as the hanuman langur, a common species in India, can also be found high in the mountains.

SPIDER MONKEY
The New World or American monkeys have a prehensile tail, which is used as a fifth limb for grasping branches.

LANGUR
Langurs have adapted well to human habitation and raid garbage dumps and picnic sites for food.

LEMURS

Lemurs are found only in Madagascar. These monkeylike animals are mostly nocturnal and feed on fruits, seeds, flowers, and leaves, gathered as they move through the trees. They have large eyes and ears, pointed noses and soft, attractive fur. The ring-tailed lemur is a popular animal in wildlife parks and zoos.

TITI MONKEY
Most monkeys eat a varied diet of flowers, fruits, seeds, insects and other small animals, and bird and reptile eggs.

Lemurs, bushbabies, monkeys, and apes (Primates)

The group includes:

Old World monkeys (Africa and Asia)
82 species

New World or American monkeys
30 species

Marmosets and tamarins
21 species

Lemurs
22 species

Bushbabies and lorises
10 species

Tarsiers
3 species

Aye-aye
1 species

Apes are shown in the next pages.

155

APES

The lesser apes include nine kinds of gibbons from Southeast Asia. They are incredibly agile climbers, using their long, muscular arms to swing themselves through the trees. They communicate with other members of their group by loud, musical cries which echo through their forest home.

There are four species of great apes. Best known are the chimpanzee and gorilla, from restricted areas in Africa. Chimps are clever animals and have been observed to use sticks and other tools to help themselves find food. They take a wide range of items, from nuts and fruit, to other animals such as lizards and small monkeys.

The pygmy chimp, or bonobo, is much rarer and less known than the chimp.

GORILLAS
The large senior male of a gorilla group, known as the silverback, protects the others and will charge at an attacker. But most of the time, gorillas live quiet, peaceful lives.

IN THE TREES
Full-grown chimps are as heavy as people and spend much time on the ground. Youngsters play in the branches as they learn to find food.

Gorillas are huge, muscular apes and are almost entirely vegetarian. They live in small groups in dense forest, feeding mainly on the leaves and stems of forest trees and vines. Orangutans live in the forests of Borneo and Sumatra. They are very fond of fruit.

More groups of primates:

Gibbons (lesser apes)
9 species
Includes:
- agile gibbon
- kloss gibbon
- lar gibbon
- siamang

Great apes
4 species
Includes:
- chimpanzee
- pygmy chimpanzee (bonobo)
- gorilla
- orangutan

Biggest primate
- The gorilla can grow to 6 feet (1.8 m) tall and weigh 400 pounds (180 kg).
- Humans are also included in the primate group. Evidence from fossils, evolution, and genetics shows that our closest living relatives are chimps.

157

PLANTS

Plants show a huge range of sizes and shapes, from tiny mosses tucked away in forests, and the miniature floating leaves of duckweeds, to towering trees such as redwoods, which are by far the biggest and oldest living things on Earth.

Plants live in nearly all habitats, on land and in fresh and salt water. There are about 500,000 different species. Around half of these belong to the group called flowering plants, or angiosperms. These include familiar garden flowers, bushes and trees, herbs and grasses. Other plant groups include seaweeds, mosses, liverworts, ferns, horsetails, and conifers (cone-bearing trees).

LIVING ON LIGHT

Most plants are green in color, especially their fronds or leaves, because they contain a greenish substance called chlorophyll. This substance enables green plants to trap the light energy in the sun's rays, to live. This process is called photosynthesis. Plants also take in minerals and raw materials from their surroundings. Using the trapped sunlight energy, they assemble these into new living tissue as they grow and breed.

Plants can grow wherever there is enough light, warmth, and moisture. They are at their most varied in warm, moist habitats such as tropical rainforests. Plants, in turn, provide food for herbivore animals, which are then food for carnivores. So plants are the basis of all life on Earth.

FLOWERLESS PLANTS

The simpler plants do not have flowers or seeds. They reproduce by tiny dustlike particles known as spores. (A spore usually consists of just one or a few microscopic cells, whereas a seed is a baby plant with its own food store.)

SPONGY LEAFLETS

Mosses and liverworts have no proper roots, stems, or leaves. A moss has rounded leaflets with fibres anchored to a rock or log. They are typical of damp

Spore capsule

Leaflets

Frond

Stem

MOSSES
The tiny spores develop in "swan's neck" capsules that grow on long stalks. The rest of the moss hugs the damp ground. It takes in nutrients from the water and dampness around, since it has no proper roots. Sphagnum moss forms deep beds in bogs and marshes.

160

places, although many mosses can withstand occasional drying, then absorb water like a sponge when wetted again.

Liverworts are flat and ribbonlike, growing almost flat against the ground.

Most ferns are larger, and also tend to grow in damp places. They have attractive feathery fronds and stiff stems that hold them up away from the ground. The stems contain tiny pipes that carry water and sap around the plant.

Main groups of flowerless plants include:

Seaweeds (algae)
12,000 species
• live in salt and fresh water
• tough, rubbery fronds
• no proper stems or roots

Mosses
10,000 species
• small and leafy
• grow in damp places

Liverworts
6,000 species
• small and mostly flat
• grow in damp places

Clubmosses and quillworts
1,000 species
• glossy leaves in spirals

Horsetails
15 species
• hollow, jointed stems
• umbrellalike spiky leaves

Ferns
12,000 species
• large fronds, often deeply toothed or branched
All of these plants breed by spores rather than seeds.

SEAWEEDS
Algae such as bladderwrack grow along shores. Their rubbery fronds withstand battering by waves.

161

CONIFERS

 Most conifers are trees, such as pines, firs, spruces, larches, and hemlock. This group contains the tallest and most massive plants in the world—the redwoods of the Pacific coast of North America. Conifers form large boreal forests in the cooler climates of northern North America and northern Asia, and also on hills and mountains.

BREEDING BY CONES

Conifer means "cone-bearer"; the male cones are usually small and produce the tree's pollen (male reproductive cells). Female cones are usually larger and woody, and contain the ovules (female cells) from which the seeds develop.

KAURI PINE
Kauris come mainly from New Zealand and Australia. They can grow to 165 feet (50 m) high and, in some types, the trunk is more than 20 feet (6 m) across. The resin or sap is kauri gum, also called copal, used for varnishes, inks, and paints.

162

CEDARS

The Lebanese cedar has several crowns of spreading branches and massively thick main boughs. Cedars are planted in many other areas as landscape trees.

PINES

Pines have tall, pointed shapes. This helps snow to slide off easily, rather than settle and snap the branches.

SIMPLE LEAVES

The leaves of conifers are simple and needle-like or scale-like. Most conifers are evergreen, keeping a full set of leaves throughout the year. In fact, the needles fall off regularly, but are continuously replaced.

Coniferous and similar plants include:

Cycads

185 species

• small shrubs; some palmlike
• bear cones
• reproduce by seeds

Ginkgo

1 species

• maidenhair tree or gingko
• fan-shaped leaves
• fleshy seeds
• reproduce by seeds

Conifers

550 species

• bear cones
• most have needle-shaped or scale-like leaves
• reproduce by seeds

Gnetophytes

70 species

• bear cones
• reproduce by seeds
• include the strange welwitschia plant of southern African deserts

163

GRASSES AND PALMS

The biggest group of plants is the flowering plants, which produce flowers, blooms, or blossoms. They are divided into two large subgroups. One is the monocotyledons, or monocots for short. This name means that when the seeds grow or germinate, they produce a single first leaf. (The other group, dicots, is shown on the following pages.) Monocots include lilies, orchids, irises, daffodils, grasses, rushes, sedges, pondweeds, duckweeds, pineapples, bananas, palm trees, and aroids.

Lady's slipper orchid

GARDEN AND FARM
Grasses are found on lawns, fields and meadows, and also as crops—wheat, barley, rice, corn, and oats. Some grasses have woody stems and grow tall, such as sugarcane and bamboo.

BAMBOO
There are more than 1,000 kinds of this tall, woody-stemmed grass. Some reach 100 feet (30 m), with stems a foot (30 cm) across. They may live to 100 years old, yet only flower and produce seeds once in a lifetime.

164

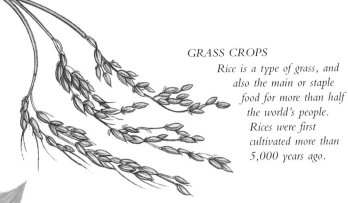

Rice is a type of grass, and also the main or staple food for more than half the world's people. Rices were first cultivated more than 5,000 years ago.

PALM TREES

Palm trees look very different from most other trees. The trunk of a palm tree is made up of tough, tightly packed leaf bases, like woody rings piled one on another. The large, fan-shaped leaves or fronds spread out at the top like an umbrella.

DATE PALM

Many palms produce useful products, such as dates, coconuts, and palm oil.

Flowering plants (angiosperms)

250,000 species

- complex plants with roots, stems, shoots, leaves
- bear flowers
- reproduce by seeds, which develop from pollinated flowers
- seeds may be contained in fruits or nuts

The two main groups of flowering plants are:

Monocotyledons

70,000 species

- seedling has one first leaf
- flower parts usually in 3s
- veins or ribs in the leaves are parallel rather than branching
- includes grasses and palms

Dicotyledons (shown on next four pages)

165

FLOWERS AND HERBS

The largest group of flowering plants are known as the dicotyledons, or dicots for short. Their seedlings produce a pair of first leaves when they grow. They have more complicated flowers than monocots, with the parts of the flowers (such as the petals) arranged in fours or fives.

This group contains most trees (except the conifers), such as beech, oak, elm, and birch, and also most flowers, herbs, spices, and vegetables, including violets, poppies, water lilies, buttercups, daisies, honeysuckle, primroses, nettles, saxifrages, cabbages, gentians, nightshades, cactuses, carnations, parsley, potatoes, and pumpkins (to name a tiny proportion).

FLOWER TO SEED

The flower is the part of the plant designed for reproduction, to make seeds. Many flowers are

Flower has red petals

New flower in bud

Seed capsule

POPPY

Poppies are sometimes called "weeds" because they grow in freshly disturbed ground. However, weeds are not a proper plant group. They are simply plants growing where they are not wanted.

bright and showy, with a distinctive scent. They attract animals such as insects to carry their pollen from the male flower parts to the female ones. The flowers offer sweet, sugary, energy-packed nectar in return.

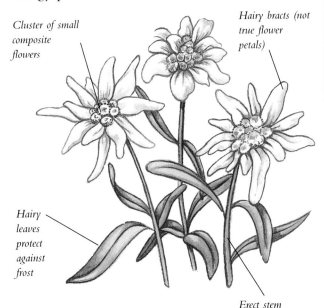

Cluster of small composite flowers

Hairy bracts (not true flower petals)

Hairy leaves protect against frost

Erect stem

EDELWEISS
Each kind of flower prefers a certain type of soil, and a certain range of rainfall, temperatures and exposure to wind, sun, and frost. The edelweiss can survive the biting winds, sharp frosts and snow cover at high altitudes in the Alps of Europe.

Largest of all plant groups, and a subgroup of flowering plants (angiosperms):

Dicotyledons
180,000 species
- seedling has two first leaves as it grows from its seed
- flower parts usually in 4s or 5s
- branched rather than parallel veins or ribs in the leaves
- includes most common flowers, herbs, bushes, and trees

- Flowering plants first appeared on Earth more than 100 million years ago, during the time of the dinosaurs. Until then, the landscape lacked their colorful petals and sweet scents.
- Grasses appeared only about 25 million years ago.

BUSHES AND TREES

Trees are tall, woody plants that grow upright, supported by a thick woody main stem, the trunk. A bush or shrub is like a tree but smaller, usually less than 16 feet (5 m) tall, and has several main trunks or stems, or one stem that branches at the base.

Most trees live for many years, even centuries. They form huge woods and forests around the world, and have been planted in gardens and parks, as windbreaks in open fields, and in plantations for timber and other products.

BROAD LEAVES
Trees in the flowering plant group have wide, flattened leaves and are called broad-leaved trees (unlike conifers with their narrow leaves). Typical broad-leaved trees include beech, birch, poplar, willow, and hundreds of other kinds.

Eucalyptus or gum tree

Mimosa (type of acacia or wattle)

168

LEAF FALL

Many such trees are deciduous. This means they lose all their leaves around the same time, usually at the start of the cold season. In some tropical areas with wet and dry seasons, deciduous trees lose their leaves at the start of the dry season. Some species of gum (eucalyptus) and oaks are evergreen, shedding a few leaves regularly through the year.

Bottlebrush

MAPLES

The maples have distinctive deeply notched leaves and winged seeds that twirl away in the wind. In most deciduous trees, the leaves become bright red, gold, and yellow in autumn. This is when the tree transfers nutrients from its leaves into its branches and trunk, before the leaves fall.

Tree records

- The coast or Californian redwood attains the greatest average height of any tree species, with some reaching more than 360 feet (110 m).
- Some species of gum trees or eucalypts approach this height. The Australian mountain ash (a type of eucalyptus) reaches 330 feet (100 m).
- The Sierran redwood (Wellingtonia or Big tree) has the largest diameter trunk of any tree—more than 30 feet (9m) across.
- The balsa or down tree has the softest and lightest wood of all, used for making model aircraft.
- Among the heaviest woods is ebony. The true ebony tree grows in India and Sri Lanka. Its black heartwood is so heavy that it sinks in water.

FUNGI AND MICROLIFE

In the past, all living things were put into one of two great groups or kingdoms—plants or animals. Modern science now recognizes three other kingdoms too. Besides animals and plants, these are monerans, protists, and fungi.

Monerans are the smallest and simplest life forms. Each is just a single microscopic cell. They include more than 10,000 kinds of bacteria, and also the blue-green algae that form the "scum" on stagnant ponds.

MOST ABUNDANT OF ALL

Bacteria live almost everywhere, inside glaciers and boiling hot springs, even inside rocks and floating in the air above mountains. We cannot see them without a microscope, but they are the most abundant living things on Earth.

Protists are also single cells. They live almost everywhere there is water, from puddles to the ocean depths. Some are like tiny animals and eat food, others are like tiny plants and capture light energy. They are described in this section, along with the fifth great kingdom, the fungi—mushrooms, toadstools, molds, yeasts, and similar life forms.

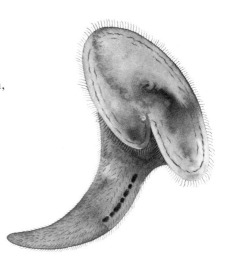

FUNGI

Fungi include not only mushrooms and toadstools, but also a whole host of other life forms. Many are not obvious, either because they are microscopic, or because they grow underground, in rotting wood, under stones or in similar out-of-the-way places.

HARMFUL AND USEFUL

Some diseases are caused by fungi, such as Dutch elm disease in elm trees, ringworm in animals, and athlete's foot in people. But other fungi are useful. Yeast is a microscopic fungus that we use to make bread rise, grape juice ferment into wine, and grains ferment into beer.

For most of its life, a fungus such as a mushroom exists only as a mass of thin fibers,

MUSHROOMS

Mushrooms and other fungi play a vital role in nature. Along with bacteria, they help to rot down or decay the parts and bodies of plants, animals, and other once-living things. In this way, they recycle nutrients and minerals back into the soil, for new growth.

Young
toadstool

Stalk

hyphae, growing into and through another living or recently dead organism. This is how fungi get their nutrients—by causing other living or just dead things to decay. A fungus becomes much more noticeable when it develops its fruiting bodies—the mushroom or toadstool itself.

Cap

Gills on underside of cap

TOADSTOOLS

Some fungi have bright colors, like the fly agaric, which warn that their flesh is poisonous or even deadly. However, some drab brown or gray fungi, such as the death's cap, can also be exceedingly dangerous. Wild fungi should never be eaten or even handled unless properly identified.

The fruiting body reproduces the fungus by releasing millions of tiny particles called spores. These blow in the wind and grow into new fungi elsewhere.

Fungi

Estimated 1.5 million types (about 60,000 so far described)

- get their nutrients by the decay or rotting of living or once-living matter
- exist mainly as a network or mycelium of tiny threads called hyphae
- reproduce by means of fruiting bodies that release spores
- fruiting bodies are obvious in types such as mushrooms, toadstools, brackets, and pin molds
- fungi also include yeasts, mildew, rusts, and blights

173

MICROSCOPIC WILDLIFE

Protists are very small, single-celled life-forms. Most are too small to see without a microscope, and go unnoticed by the naked eye. Some kinds, called protozoa, are each like a single animal. They can move about and feed by eating even tinier organisms such as bacteria. Others, such as euglenas, are more like plants. They can make their own food, using the sun's light energy.

Food particles

WHERE DO THEY LIVE?

Protists are found in most places where there is water or moisture—in the sea, in ponds, in the soil, and also on and inside animals and plants. Many are free-living, but others can only survive as parasites inside animals or plants. Some cause diseases,

Control center (nucleus)

STENTOR
This is a large protist, as big as a pinhead. It is funnel-shaped and is normally attached by its base. However, it can also swim freely in the water, propelled by beating its tiny hairs or cilia.

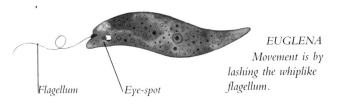

EUGLENA
*Movement is by
lashing the whiplike
flagellum.*

Flagellum   Eye-spot

AMOEBA
*This bloblike protist can move about by changing its
shape. It feeds by surrounding particles of food with its
own body and engulfing them. The undigested
remains are left behind as the amoeba oozes
on. Some amoebas are real giants, growing
to almost 0.4 inch (1 cm) across.*

PARAMECIUM
*The ciliate protists are
covered by tiny hairlike
growths called cilia. These
beat in waves to move the
Paramecium along. There is
also a "mouth" for eating tiny
food particles.*

Covering of cilia

Mouth (digestive groove)

Excess water collects and is expelled

Protists
*100,000 types (with many more
to be identified)*

- single-celled body
- most kinds are microscopic
- some engulf or eat tiny
  food particles
- others capture the sun's
  light energy by
  photosynthesis
- still others do both of these,
  together or at different
  times
- single-celled algae (simple
  green plants) are sometimes
  included here
- live almost everywhere,
  especially in the sea, fresh
  water and the soil
- reproduce by simply
  splitting or dividing into
  two (binary fission)
- some also reproduce by
  spores

Protists provide food for
slightly larger animals and so
form the first parts of many
food chains and food webs.

175

WILDLIFE HABITATS

Different areas of the world have different climates and soils, and so support their own characteristic wildlife habitats, where certain plants and animals thrive. These vary from the cold, icy, and snowy conditions of the polar regions, to the lush forests of the always-warm tropics.

Between these extremes, temperate regions have different kinds of forests, such as conifers or broad-leaved trees, where the climate is moist enough. In areas where the rainfall is too low or irregular to support tree growth, scrub or grassland grows. Even drier places have semideserts or deserts.

WATERY HABITATS

There are also many different watery or aquatic habitats. Freshwater types include the scattered pools of bogs, marshes, and swamps, flowing water in streams, and rivers, and the still water of ponds and lakes.

The seas also have their own particular wildlife habitats. They include the cold polar oceans, warm coastal mangrove swamps, rocky and sandy shores, sunlit coral reefs, the wide-open surface waters of the oceans, and the biggest habitat of all, but the one least known to us—the vast, dark depths of the sea.

POLAR AND MOUNTAIN

The regions surrounding the North and South Poles are cold and windswept, and in winter they are also dark. Much of these areas are covered in ice and snow, and few animals and plants live here. But the lands around the Arctic are slightly warmer in summer, and the ice melts to form the tundra habitat. Sedges and grasses form dense tussocks. Flowers grow in low mats or firm cushions, away from the drying winds. Many produce large, bright blooms. Millions of flying insects breed in the endless maze of pools.

ARCTIC CREATURES
Arctic mammals include lemmings and voles, polar bears, Arctic foxes, wolverines, Arctic hares, and reindeer (caribou). In the sea, there are walruses and harp, hooded and ringed

DAILY ROUTINE
Mountain animals, such as the mountain goat, move up the slopes in the morning warmth to nibble plants. As cold night falls, they return to the shelter of the trees on the lower slopes. Many mountain birds follow the same routine.

SEASONAL MIGRATION
Birds and large mammals travel or migrate north in the Arctic spring, to take advantage of the brief but rich summer growth. They return south each autumn.

seals, as well as whales such as the killer whale, beluga, and narwhal.

Polar bear

The southern continent of Antarctica is frozen almost all year round. Only a few hardy plants, such as mosses, and some types of tiny insects can survive.

A mountain shows the same series of habitats going up in altitude, as seen when traveling from the tropics to the poles

- Surrounding lowland plains may be grassland or scrub
- Foothills of broad-leaved trees
- Lower alpine shrubland
- Upper forests of conifers
- Higher alpine meadows above the tree line
- Tundralike icefields near the summit
- Permanent ice and snow on the peak

179

CONIFER FORESTS

The boreal forest or taiga is the largest forest in the world, stretching for 7,000 miles (12,000 km) through Siberia, Alaska, Canada, Scandinavia, Finland, and northwest Russia. The largest tracts of taiga are in Siberia, where they may be as much as 1,200 miles (2,000 km) wide north to south, and in Canada, where the forest is about 550 miles (900 km) in width. The taiga stretches from Labrador across Canada to Alaska. Most of the taiga is dark, coniferous forest, broken up every now and then by rivers, bogs, swamps, or lakes. At its northern limit, the forest gradually thins out, and the trees are smaller and sometimes stunted, giving way eventually to the treeless, boggy plains of the tundra. Light levels and temperatures are low for most of

TREE TYPES

The main trees of the taiga are spruce, fir, pine, and larch, although in the more open sites some broad-leaved trees such as birch, willow, alder, and aspen also grow. The trees are well adapted to the snowy conditions. Their tentlike shapes, with branches sloping downward, allow the snow to slide off, and their thick bark offers protection from extreme cold.

WAXWING
This common bird is named from the red, droplike, waxy feather shafts visible on its wings. It eats seeds and berries.

WOLVERINE
The largest member of the weasel and stoat family, the wolverine is a powerful predator. It also scavenges on large animals such as caribou that succumb to the cold.

the year, and this kind of forest contains relatively few species. On the forest-floor, the vegetation is sparse, with low shrubs, and many lichens, mosses, and ferns. The needle-like leaves of the trees rot very slowly in the cold conditions and form a thick layer of leaf litter.

Larger mammals of the northern forests include:

- Brown (grizzly) bear
- Moose
- Caribou, which is the common name in North America for the animal known as the reindeer in Northern Europe
- Gray wolf
- Lynx
- Wolverine
- Sable

The most common rodents are:

- Voles
- Lemmings, which occur throughout, often in very large numbers

Typical birds of the taiga are:

- Great gray owl
- Crossbill
- Waxwing
- Goshawk
- Three-toed woodpecker

181

DECIDUOUS FORESTS

Broad-leaved or deciduous forest is the natural vegetation in the hills and lowlands of the temperate regions, for instance, in much of eastern North America, western Europe, and in eastern Asia. It tends to develop in areas of warm, moist summers, where the annual rainfall is 20 inches (500 mm) or more, and where the winters are cool.

Unlike the conifer forests or taiga, much original broad-leaved forest has been cleared for farmland and human habitation. The areas remaining are tiny remnants of the original tracts.

FERTILE SOILS

The soils of broad-leaved forests are highly fertile, because they have developed over centuries of regular leaf fall and decay. This makes them very

LEAF-FALL

"Deciduous" means that a tree loses its leaves all at one time, usually at the onset of the cold or dry season. The leaves turn from green to brown, yellow and gold, as their nutrients and pigments are broken down and absorbed into the tree's branches and trunk. The leaf comes away and the leaf scar seals over to keep out frost.

suitable for growing crops after they have been cleared. But the soils must then receive large amounts of fertilizers, to retain their goodness.

Since most of the trees drop their leaves in the autumn, plenty of light enters these forests during winter and spring. Many plants of the woodland floor are adapted to take advantage of this by growing and flowering in early spring, before the new leaves sprout. Many deciduous woods are carpeted with bright spring flowers, such as anemones and bluebells.

LARGEST DEER

The moose, also called the elk, lives in deciduous and conifer forests around the Northern Hemisphere. It has also been introduced to New Zealand. It wades into water to feed on soft aquatic plants, and is an excellent swimmer. In winter the moose gnaws on woody bark, shoots, and other plant parts.

Deciduous woodlands support many animals:

- Woodpeckers probe into tree bark for grubs
- Squirrels feed on seeds, berries, and fungi
- The sparrowhawk is a specialist woodland predator, using its rapid, swerving flight to ambush small birds in the air
- Jays are mainly found in woodland, especially where there are plenty of oaks

Typical trees of these forests:

- Oaks
- Lime
- Maples
- Elm
- Beech
- The warmer forests towards the southern edge of the region are much richer, with other species, including (in North America) tulip tree and hickory

183

TROPICAL FORESTS

Tropical forest is the natural vegetation of much of the lowlands in the moist tropical regions of the world.

However, huge areas of this type of forest have been felled or damaged by people, although some tracts are still found, especially in South and Central America, and in equatorial Africa. Almost one third of tropical forests are in Southeast Asia, where they are found scattered from India and Sri Lanka, to Thailand, Vietnam, and Cambodia, to Indonesia, and across to eastern Queensland in Australia. The biggest area of tropical forest lies in the basin of the mighty Amazon River in Brazil, and also along the foothills of the Andes mountains. These forests also stretch northward to Panama, Costa Rica, Honduras, Guatemala, and Belize.

HARPY EAGLE
This eagle has the largest body of any eagle in the world, but it also has relatively short wings, compared to other types of eagles. This allows it to hunt among the rainforest trees, rather than soar high in the sky. It lives in South America and feeds on monkeys, sloths, opossums, and birds.

184

The wet tropical forests, often called rainforests, are the most complex of all natural communities, with a huge wealth of plant and animal species. Here the tall trees, mostly evergreens, tower to 100 ft (30 m) or more. In lowland tropical rain forest, the mean annual temperature usually exceeds 80°F (27°C), and rainfall may be more than 80 inches (2,000 mm) per year.

TROPICAL EAGLE OWL

GABOON VIPER
The zigzag pattern of browns makes the Gaboon viper very well camouflaged among the dead leaves of the forest floor.

Nature's greatest diversity :

- Tropical forests contain the greatest diversity of plants and animals of any habitat.

- There are as many as 200 different species of trees alone in 2.5 acres (1 ha).

- Botanists estimate that these forests may contain more than 100,000 plant species— 40 percent of the world's total.

- The animal life is equally rich with the largest concentrations of mammals and birds, and many new species (especially insects) discovered each year.

- Tropical forests are, on average, at least one hundred times more diverse (in numbers of species) than temperate deciduous forests.

185

GRASSLANDS

Where the climate is too dry for trees to grow, or the soil is too poor in nutrients, grassland tends to be the dominant habitat in temperate regions. It develops mostly in areas where the summers are hot, but the winters cold. The soils of temperate grasslands are deep and fertile, ideal for cultivation of crops, especially members of the grass family such as cereal crops, notably corn and wheat. For this reason, huge swathes have been plowed for crops, or grazed by domestic animals.

GRASSLAND REGIONS
The main regions of natural temperate grassland are the central prairies of North America and the steppes of Asia, with smaller areas in the pampas of Argentina, the veld in

SECRETARY BIRD

Long, strong legs allow the secretary bird of Africa to walk up to 20 miles (32 km) daily, and stamp to death its prey of small mammals, birds, and even reptiles such as lizards, and poisonous snakes.

South Africa, and the grasslands of the southeast of South Island in New Zealand. The prairies and steppes are the largest areas of original grassland, covering huge chunks of North America and Asia.

TROPICAL GRASSLAND

The tropical grassy plains of Africa are called savanna. This is a mosaic of habitats, with patches of grassland alternating with thickets of trees and scrub. Most grassland animals, such as gazelles and zebras, are fast runners, able to flee from predators, or small burrowers that dig tunnels for safety. There are few trees for nesting, so many birds are ground-nesters and good runners too.

Mammals of the North American prairies:

- pronghorn (antelope)
- Rabbits
- Coyotes
- Bobcats
- Badgers
- Ground squirrels
- Marmots
- Gophers
- Snakes

Typical prairie birds:

- Larks
- Buntings
- Grouse
- Prairie chickens
- Turkey vultures

On the Eurasian steppe:

- Wild horses once grazed, but are now extinct as a truly wild variety
- The saiga antelope is another specialist grazing mammal
- Other steppe species are susliks, marmots and steppe polecat

187

DESERTS

Deserts cover about one fifth of the land surface of the Earth. They form where the rainfall is so scarce (less than 10 inches/250 mm per year) that the soil remains very dry for most of the time, usually because the area is in the middle of a continetal land mass.

Some deserts receive no rain for years on end. They may be sandy, like much of the Sahara, or rocky, or a mixture of the two.

COASTAL DESERTS

Deserts also form in areas where offshore winds predominate, as in the Atacama Desert of Chile and the Namib Desert of West Africa. Here the air is nearly always dry, the winds losing most of their moisture as they pass over the mass of each adjacent continent.

DESERT EROSION

Some desert areas are covered with loose sand, such as parts of the Sahara in Africa and the Great Sandy Desert in Australia. But more than one third of deserts are rocky, strewn with pebbles and loose stones. These heat up in the fierce midday sun, then cool rapidly at night. The continued temperature changes make the rock crack and flake, breaking off chunks that roll around, break further and finally blow away in the strong winds.

The main arid (dry) areas and deserts are the Sahara of North Africa, the Namib and Kalahari also in Africa, the Gobi of Asia, the Great Australian Deserts, the Atacama in Chile, the Great Basin of North America, and the Sonoran, Mojave, and Chihuahuan Deserts of southwest United States and Mexico.

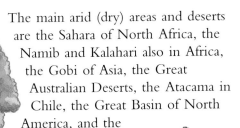

Baobab tree

Saguaro cactus

FENNEC FOX

The fennec fox is a dainty carnivore living in the sandy deserts of North Africa and Arabia, and is mainly nocturnal. It uses its large ears both to help locate insects and other small animal prey, and also as radiators to help it lose heat.

How plants survive in the desert:

• Some plants survive the harsh conditions as seeds, germinating, growing and coming into flower rapidly after the occasional heavy storm.

• Other desert plants persist as bulbs below ground, growing up only after rain.

• Succulent plants, such as the cacti of America and the euphorbias of Africa, often have very thick cuticles, with sharp spines. These adaptations prevent excess water loss and reduce damage from grazing animals.

• Many plants can also store water in their spongy cells.

189

MARSHES AND SWAMPS

Marshes and swamps tend to develop in low-lying ground alongside rivers, especially if the water flow has been impeded by a barrier such as a sand bar or spit. Large river deltas can be an intricate network of channels and marshy habitats. These are a haven for wildlife of all sorts, most notably for birds, fish, and amphibians. Many snakes can also swim well and find wetlands profitable hunting grounds.

WETLAND BIRDS

Herons and their relatives are some of birds best suited to life in and around a marsh. They wait patiently in the shallow water for their prey—mainly fish and frogs. Many also nest in colonies, and the swampy

COTTONMOUTH
A very poisonous snake, the cottonmouth or water-moccasin is active at night and hunts fish, frogs, small mammals and other reptiles. Unusual for a snake, it also eats carrion.

Brown pelican

ground gives them some protection from land animals. The reedbeds which dominate numerous marshes are home to many other birds, including the camouflaged bittern, and reed and sedge warblers that hang their nests between the reed stems.

Osprey (seahawk or sea eagle)

AERIAL ROOTS

The swampy ground of wetlands often contains little oxygen. Aerial roots can obtain oxygen from the air and also stabilize the tree in the soft mud.

Some famous wetlands:

- Everglades, Florida
- Volga Delta, Russia
- Camargue, southern France
- Danube Delta, Romania
- Coto Donana, southern Spain
- Okavango Delta, Botswana, Africa

Mammals of marshes and swamps include:

- Beavers
- Muskrats
- Otters
- Mink

Insect life:

The insect life of wetlands is extremely rich, with many kinds of aquatic larvae, from mosquitoes to dragonflies

- As well as feeding fish, these larvae provide food for thousands of waterfowl such as ducks

191

RIVERS AND LAKES

Lakes are found throughout the world, differing in size, depth and water chemistry, and providing key habitats for wildlife. Ducks are prominent amongst the bird life, either dabbling near the surface, like mallards, or diving below, like tufted ducks and pochards. Grebes construct their

floating nests on the water, using reeds and other vegetation. They feed mainly on invertebrates and fish. Some birds, such as kingfishers, catch fish by diving from an overhanging branch.

CLINGING ON

The wildlife of rivers has adaptations to prevent being swept away by the current. Insect larvae cling firmly to stones on the river bed. Some river fish, such as the bullhead, shelter in the lee of rocks or boulders, but river fish of open water, such as trout,

ABOVE AND BELOW
Birds such as herons (left) stalk their prey slowly along the riverbank or in the lake shallows. The victim may be a crayfish (right), a freshwater relative of the lobster.

Tail fan (telson)

large scaly
plates (scutes)

STURGEON
This massive fish
can grow to more
than 10 feet (3 m)
long. It was hunted
heavily for its eggs (roe),
known as caviare, and
is now very rare.

must swim constantly to maintain position.
Only plants firmly rooted in the bed can
survive in faster-flowing rivers. However, the
weaker flow near the bank, allows a fringing
growth of reeds, sedges, and other plants.

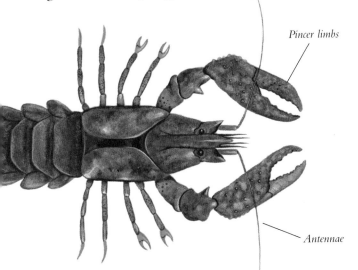

Pincer limbs

Antennae

Lake records:

- The largest lake, with a
  surface area of 142,145
  square miles (371,000 sq
  km), is the Caspian Sea,
  bounded mainly by Russia
  and Iran. It is salty, and is
  probably the land locked
  remnant of a former sea.
- The largest freshwater lake,
  with a surface area of about
  31,500 square miles (82,260
  sq km), is Lake Superior.
- Lake Baikal in Siberia is the
  world's deepest lake, with a
  maximum depth of 5,311
  feet (1,620 m) and averaging
  about 2,400 feet (740 m).
  This one lake holds 20 per
  cent of the world's
  freshwater, and it is fed by
  330 rivers.
- Baikal is home to the only
  freshwater seal, which feeds
  mainly on deepwater fish
  and gives birth in spring in
  lairs dug into the snow on
  the frozen lake.

193

COASTS

Coastal habitats range from flat expanses of open mud, to cliffs, rocky shores, and sandy bays. The ebb and flow of the tide twice each day brings many benefits, such as well-aerated water and fresh supplies of food from the sea. However, it also poses problems for coastal wildlife, which must adapt its activities accordingly.

ON THE ROCKS

On a rocky shore, the seaweeds, and animals such as mussels, limpets, barnacles, and sea anemones, are anchored firmly to the rocks, otherwise they would be swept away. The animals also need to open up or emerge from their shells for feeding when the tide is in. Then they close up tightly for protection against predators and drying in the sun, or exposure to fresh water as rain, when the tide is out.

SWIFT ROCK CRAB
Crabs breathe by feathery gills, which work only in water. When they emerge into the air, they carry a small volume of water inside their gill chambers, within the shell. They can breathe for a short time using this. But they return regularly to water to refresh the supply.

Different kinds of seaweed grow at different levels on the shore. Bladderwracks are mostly near the high water mark. Kelps are restricted to the low-water mark or below, where they remain submerged most of the time.

Red seaweeds grow in shallow water, and cannot withstand being exposed to air for any length of time.

ANEMONES

These simple animals have stinging tentacles to catch their prey of small fish, prawns, free-swimming worms, and similar marine creatures. The prey is taken into the mouth in the center of the tentacles, and slowly digested. The leftover remains come out of the same mouth. Anemones live mainly anchored to rocks, although some attach themselves to large fronds of kelp.

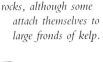

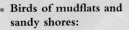

Birds of mudflats and sandy shores:

- A retreating tide on a flat muddy or sandy shore reveals vast areas of feeding ground to large numbers of wading birds, such as dunlins, sandpipers, and oystercatchers.
- These probe into the soft soil for small crustaceans, worms, and other invertebrates.

Birds of coastal cliffs:

- Seaside cliffs are usually colonized by birds because they are relatively safe breeding sites, out of reach of predators such as foxes, lizards, and rats.
- As well as gulls, other typical cliff-nesting birds are auks such as guillemots, razorbills, and puffins, and cormorants and shags.

SEAS AND OCEANS

On land, the basis of the food chain is easily visible—plants such as grasses and trees. Ocean life similarly depends on plants, but they are less obvious—the tiny floating plants of the plankton. These grow using the sun's energy, like land plants, and nutrients from the water around them. They form the base of the ocean food chains. Tiny floating animals, making up the zooplankton, feed on the smaller plant plankton. Both provide food for slightly larger animals, such as shrimps,

CORAL REEF
Coral reefs grow only in warm, clear, shallow tropical seas, either close to the coast, or fringing islands. Coral reefs are incredibly rich in wildlife, with a mass of invertebrates such as starfish and anemones, and seemingly endless shoals of often brightly colored fish. Moray eels lurk in crevices in the coral.

BIGGEST PREDATOR

The sperm whale, weighing perhaps 50 tons, is by far the largest hunting animal on Earth. It dives hundreds of metres in search of large victims, especially giant squid and deep-sea fish. It can stay underwater for more than one hour.

The biggest habitat:

- Seas and oceans cover almost three quarters of the surface of the planet.
- They provide a rich habitat, inhabited by a multitude of animals, from huge whales to the tiniest of microscopic creatures.
- There are two main habitats in the oceans.
- One is the open water itself, sometimes called the pelagic habitat.
- The other is the ocean floor, or benthic habitat.

Major coral reefs:

- Great Barrier Reef, Australia
- Islands of Indian Ocean
- Islands of Caribbean
- Islands of Pacific Ocean
- Islands of South China Sea and Indonesia

krill, and fish fry, and the larvae (young forms) of crabs, starfish, worms, and many other creatures. Next in the chain come arrow worms, squid, and fish.

OCEAN BLOOMS

In some areas, such as around Antarctica, upwelling currents bring nutrients to the surface. These fuel huge amounts or "blooms" of plankton. In turn the blooms feed larger animals, through to the biggest predators on Earth, such as sperm and killer whales. The great whales, such as blues, filter-feed on shrimplike krill.

Biology of Wildlife

From the outside, plants and animals appear in a huge variety of shapes, sizes, body proportions and colors. But inside, there are great similarities within the major groups. The blooms of most flowering plants have the same parts, such as stamens and carpels, for making pollen and growing seeds. Despite their difference in size, a blue whale and a mosquito have many of the same main parts, or organs, inside their bodies, such as heart, brain, muscles, and intestines.

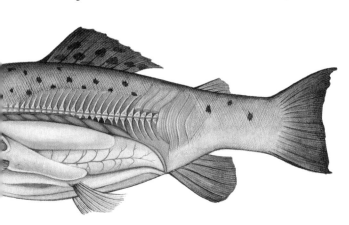

MANY BENEFITS

The study of the structure of animals and plants, and how they work inside, is part of the biology of wildlife. It helps us to understand why animals need certain kinds of food, how they move, and how they breed. In plants, these studies show us why plants need certain minerals from the soil, or how they are attacked by pests and parasites. This increases our understanding of nature, and also helps us to care for our crops and farm animals more effectively.

PLANT BIOLOGY

A flowering plant takes in water, minerals, salts, nutrients, and other simple substances through its roots. The tiny tip of each root is covered with fine fibers, thinner than human hairs, and these in turn are covered with even finer, microscopic root hairs. The result is a large surface area, 20 times the size of a root without these hairs, for absorbing raw materials from the soil. The materials pass around the plant in a network of thin tubes, xylem vessels.

LEAVES AND SAP

The leaves of a plant absorb sunlight energy and trap it by substances called photosynthetic pigments, such as green chlorophyll. The trapped energy is stored as high-energy sugars.

These sugars dissolved in water make a sweet, sticky fluid, sap. This moves around the plant in thin tubes,

phloem vessels, which often run alongside the xylem vessels. As the sap is distributed around the plant, its sugars are broken down chemically to release their energy. This is used for assembling raw materials into new plant tissues for growth, repairing wear and tear, making new roots and leaves, and producing flowers and seeds.

WOODY STEMS
The stems of bushes and trees are stiffened with woody fibres of the substance lignin. This allows the plants to grow much taller.

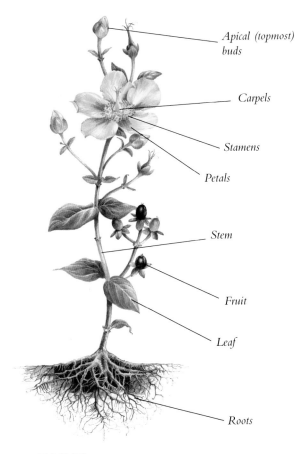

Apical (topmost) buds

Carpels

Stamens

Petals

Stem

Fruit

Leaf

Roots

PLANT PARTS
The stiff stem holds the leaves and flowers above ground, away from herbivorous animals and trampling damage, and nearer the sunlight.

Flower parts:

Stamens

These are the male reproductive parts of the flower. They consist of:

- A stalk, or filament, that holds up the anther.
- The anther, which is like a two-part bag or capsule containing pollen grains.
- Pollen grains, like tiny pieces of dust, which contain the male cells.

Carpels

These are the female reproductive parts of the flower. They consist of:

- A stigma, which is a small swelling or "landing pad" specialized to receive pollen grains.
- The style, which may be long and stalk-like, and holds up the stigma.
- The ovary, which is at the base of the style, usually in the middle of the flower.
- The ovules or female cells, inside the ovary.

INSIDE AN INSECT

An insect's body is enclosed in a hard outer casing, the exoskeleton. This is very thin at the joints, to allow bending. A worker ant shows all the typical insect organs, except the reproductive parts, since worker ants are sterile (only the queen produces and lays eggs). Breathing is by a network of air tubes, trachea, through the body. These open at small holes called spiracles on the surface.

HEART AND BLOOD

An insect's circulatory system is "open." Blood, more correctly called hemolymph, is pumped by the heart along large blood vessels at the top of the body. The hemolymph oozes out into the general body cavity or hemocoel, flowing past and bathing the various inner organs and parts. Then it collects in a series of funnel-like veins and is channeled back to the heart again.

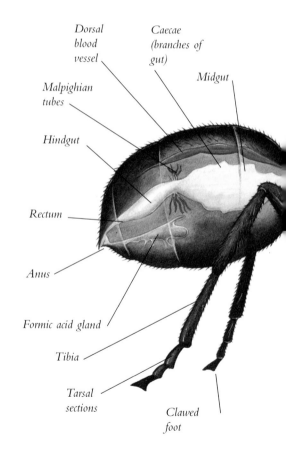

Dorsal blood vessel

Caecae (branches of gut)

Midgut

Malpighian tubes

Hindgut

Rectum

Anus

Formic acid gland

Tibia

Tarsal sections

Clawed foot

Brain

Antenna

Foregut

Heart

Compound eye

Optic nerve

Oral opening (mouth)

Mouthparts

INSIDE AN ANT
The head contains the main
sense organs, mouthparts and
brain. The thorax is mostly
muscles that move the wings
and, in winged ants at
breeding time. The thorax is
mostly digestive and excretory
organs.

Parts of the animal body:

- Abdomen—region of the body behind the thorax
- Alimentary canal—the gut or digestive tube between an animal's mouth and anus
- Aorta—the main blood vessel that leaves the heart
- Axon—the long process or fibre of a nerve cell that carries impulses on to the next cell
- Bladder—muscular bag where the waste liquid urine from the kidneys is stored before being eliminated from the body
- Brain—organ in the head of an animal where sensory nerve messages are received and sorted and motor messages sent out to the muscles in response

(Continued on page 205...)

INSIDE A FISH

A fish's body is mostly muscles, arranged in a zigzag pattern along the sides of the vertebral column, or backbone. The digestive, reproductive and other organs are packed into the lower one-third of the body. Most fish have a swim bladder, which can be filled with bubbles of gas to help the fish rise or float still in the water. However, sharks and rays lack a swim bladder. They must keep swimming to stay in midwater.

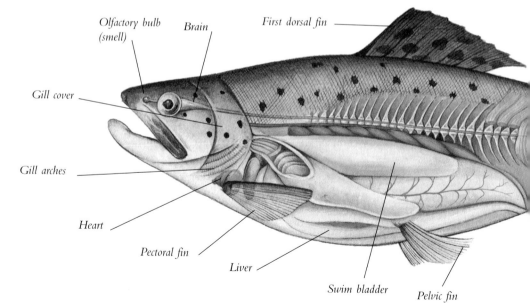

Olfactory bulb (smell)

Brain

First dorsal fin

Gill cover

Gill arches

Heart

Pectoral fin

Liver

Swim bladder

Pelvic fin

BLOCKS OF MUSCLE

The fish's muscles are arranged in blocks, called myotomes. This is an evolutionary leftover from the ancient ancestors of fish, which had body segments, like worms. The individual bones of the spinal column have long rodlike extensions for anchoring these muscles. The myotomes pull on each side alternately, bending the backbone to make the tail swish and propel the fish forwards.

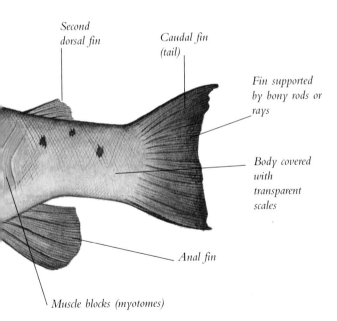

Second dorsal fin

Caudal fin (tail)

Fin supported by bony rods or rays

Body covered with transparent scales

Anal fin

Muscle blocks (myotomes)

Parts of the animal body continued:

- Cloaca—pouch that opens to the outside into which the gut, bladder, and reproductive organs empty their products
- Compound eye—a collection of several light-sensitive units (ommatidia) that make up the arthropod eye
- Crop—enlargement of the gut before the stomach where food is stored before digestion begins
- Dermis—the outer skin
- Endocrine gland—a ductless gland that secretes its product (usually hormonal) directly into the blood stream
- Gills—respiratory organs of aquatic animals where blood vessels are brought close to the skin surface

(Continued on page 207...)

INSIDE A REPTILE

A reptile is a quadruped vertebrate—it has four limbs and a bony skeleton with a backbone (vertebral column). However, during evolution, some reptiles have changed their body plan. Snakes, for example, have lost their limbs.

HEART AND LUNGS

Unlike fish, a reptile breathes air using its lungs. Blood flows through the lungs and gathers oxygen from the air inside them. Then it goes back to the heart and is pumped around the body, distributing oxygen and nutrients to all the tissues and organs, before returning to the heart.

The lungs circuit is known as the pulmonary circulation, and the body circuit is the systemic circulation. However, the two circuits are not quite separate. Blood mixes between the two in the heart. In birds and mammals, the circuits are separate.

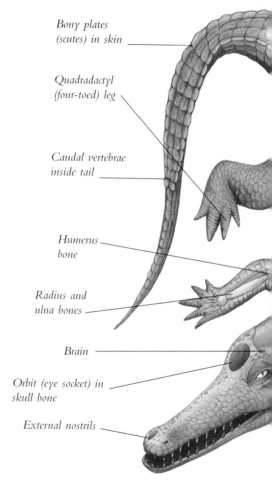

Bony plates (scutes) in skin

Quadradactyl (four-toed) leg

Caudal vertebrae inside tail

Humerus bone

Radius and ulna bones

Brain

Orbit (eye socket) in skull bone

External nostrils

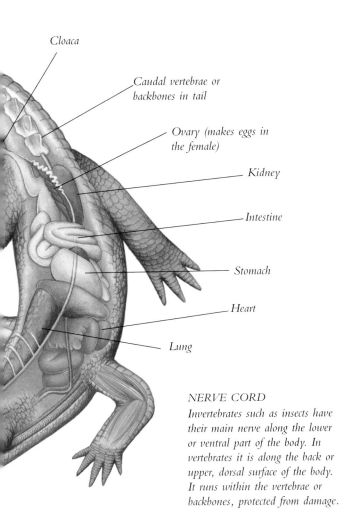

Cloaca

Caudal vertebrae or backbones in tail

Ovary (makes eggs in the female)

Kidney

Intestine

Stomach

Heart

Lung

NERVE CORD

Invertebrates such as insects have their main nerve along the lower or ventral part of the body. In vertebrates it is along the back or upper, dorsal surface of the body. It runs within the vertebrae or backbones, protected from damage.

Parts of the animal body continued:

- Gizzard—a part of the gut with muscular walls where food is mashed up or ground with stones
- Gland—a group of cells that produces a specific substance such as a hormone or digestive juice
- Heart—muscular organ that pumps blood around the circulatory system
- Intestines—long tube where digested food and water is absorbed into the blood stream
- Kidney—organ where waste products are filtered from blood and the water content balanced.
- Liver—organ that processes digested food and regulates body chemistry

(Continued on page 209...)

INSIDE A BIRD

A bird's body is highly adapted to save weight, for ease of flying. It is covered with lightweight feathers, and its bones are hollow. It also lacks teeth, which are relatively heavy, and has a beak made of the lightweight horny substance, keratin.

BREATHING

A bird's life is extremely active and flying uses up large amounts of energy and oxygen. To obtain this oxygen, the bird's respiratory or breathing system does not take air into the lungs and then breathe it out. The air flows in a more continuous one-way direction through the lungs, using a system of hollow bags called air sacs, in various parts of the upper body and also in the bones.

This helps to save weight and also makes the bird's lungs up to five times more efficient at taking in oxygen, than our own lungs.

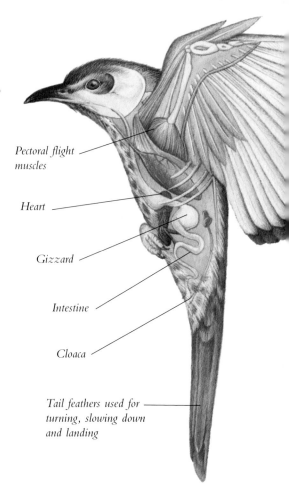

Pectoral flight muscles

Heart

Gizzard

Intestine

Cloaca

Tail feathers used for turning, slowing down and landing

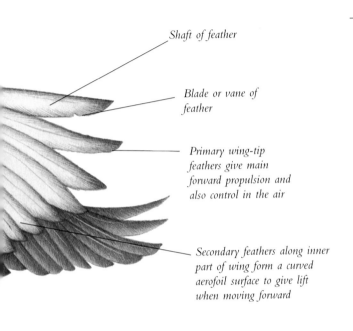

Shaft of feather

Blade or vane of feather

Primary wing-tip feathers give main forward propulsion and also control in the air

Secondary feathers along inner part of wing form a curved aerofoil surface to give lift when moving forward

FEATHERS

A bird has two main kinds of feathers. The contour feathers over its body and on its wings have large, flat, bladelike surfaces. They give protection, repel water and form an airtight surface for flying. Under the contour feathers on the body are soft, fluffy down feathers. These form an air-trapping blanket to retain body heat and keep the bird warm.

Parts of the animal body continued:

- Lung—organ where inspired air is brought into close contact with blood for the exchange of respiratory gases oxygen and carbon dioxide
- Mandible—lower jaw bone
- Olfactory organs—the organs of smell
- Operculum—the bony flap which covers the opening to the gills in fish
- Ovary—organ where eggs are produced
- Pelvic girdle—the hip, the bones that support the hind limb bones

(Continued on page 211...)

INSIDE A MAMMAL

The main internal differences between different types of mammals are in the shapes of the bones, and the muscles that pull them, and also in the size of the intestines.

Herbivores such as zebras and elephants have very large stomachs and intestines. These are filled with "friendly" bacteria that help to digest or break down the tough cellulose and other substances in their nutrient-poor plant food.

Meat-eaters such as cats, dogs, and seals have smaller stomachs and intestines because their food is highly nutritious and easily digested.

SKELETON SHAPES

Fast-running mammals such as horses and gazelles tend to have long, slim limbs that can be swung to and fro very rapidly. Heavier mammals such as rhinos, and hippos have stumpy, pillarlike legs to carry their body weight.

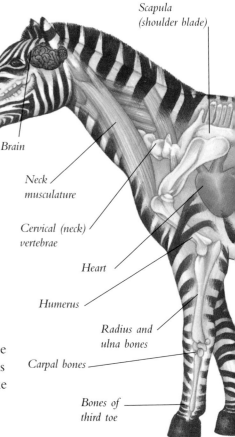

Scapula (shoulder blade)

Brain

Neck musculature

Cervical (neck) vertebrae

Heart

Humerus

Radius and ulna bones

Carpal bones

Bones of third toe

HOOFED MAMMALS

The ungulates or hoofed mammals have fewer than the usual five mammal toes on each foot. In zebras and horses there is only one toe, the third. This is large and thick-boned, a design that saves weight without losing strength, for faster running.

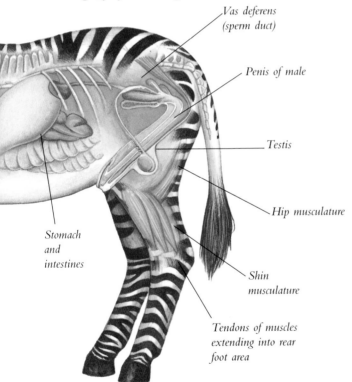

Vas deferens
(sperm duct)

Penis of male

Testis

Hip musculature

Stomach
and
intestines

Shin
musculature

Tendons of muscles
extending into rear
foot area

Parts of the animal body continued:

- Pharynx—the throat
- Stomach—part of the digestive system where food is broken down chemically by mixing with strong digestive juices
- Testes—organ where sperm are produced
- Thorax—the part of the body between the head and the abdomen, being the middle section of an insect's body or the chest area in a mammal
- Trachea—breathing tube or windpipe
- Uterus—the womb, where a baby mammal grows and develops inside its mother's body

ECOLOGY OF WILDLIFE

Animals and plants in nature do not live isolated lives, independent of one another. Instead, habitats consist of complicated networks of relationships between many different organisms, with the activities of each species affecting others, either directly or indirectly.

At a global level, we are all dependent on wildlife. Green plants trap the energy of the sun and convert this into organic material in their own bodies. In turn, they support the animal life of the planet—including the human species.

FOREST ECOLOGY

Trees and forests in particular play an important role in replenishing oxygen in the air we breathe, and absorbing carbon dioxide. As trees and other plants shed their leaves, or mature and die, so the chemicals contained in them are returned to the forest ecosystem via the soil.

This natural recycling maintains the fertility of the forest—as long as the trees remain. The last section of the book explains what happens when we endanger wildlife. We threaten the global ecosystem, too.

FOOD CHAINS AND WEBS

 The relationships between various animals and plants in their habitat can be looked at according to their feeding links. For example, a caterpillar feeding on an oak leaf is also linked to the bird it is eaten by, with the plant and animals forming a kind of natural chain along which nutrients and energy flow. In fact, most natural systems are much more complicated, and the picture we build up is more weblike than a simple chain. This is why the term "food web" is a more accurate description.

SIMPLE FOOD WEB
Four living things in a pond show simple feeding relationships. The plant produces living material. The caddisfly larva eats the plant, the diving beetle eats the caddisfly larva, and the water louse scavenges on the dead bodies of all of them.

IN THE POND
In a pond, for instance, the tiny algae and green plants growing in the water are the first link. They are the organisms that trap the sun's energy in their growing bodies, through the

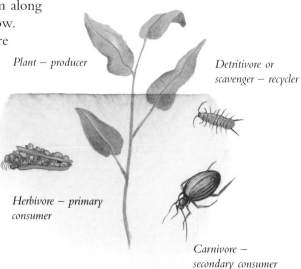

Plant – producer

Detritivore or scavenger – recycler

Herbivore – primary consumer

Carnivore – secondary consumer

214

COMPLETE HERBIVORE

The Arctic ground squirrel, or suslik, eats an almost entirely herbivorous diet. It feeds on stems, leaves, flowers, fruits, seeds, shoots and roots — but rarely eats animals.

process of photosynthesis. Many invertebrates feed on the water plants, but so do many larger animals, such as swans and ducks, and mammals such as water voles and beavers.

The invertebrates also provide food for water birds, as well as for fish and amphibians, and for mammals such as water shrews. In this way the web of feeding relationships builds up, with some animals moving from one level to another, such as herbivore to carnivore, depending on the food they are eating.

A COMPLEX WEB

If each feeding link is drawn on a picture of the pond and its inhabitants, the picture soon gets crisscrossed with lines. The incredibly complicated nature of the food web—even in a small pond—becomes clear.

Predator-prey cycles:

- Sometimes the numbers of a species build up to very high levels. This could happen during a few years when food is plentiful and the weather is mild.
- For a time, the members of the species escape from the check on their numbers normally provided by their predators or by harsh conditions.
- A well-known example is the lemming, a small arctic rodent that resembles a vole. Lemming populations rise fast to a peak every 3-6 years.
- As this happens their predators, such as arctic foxes, weasels, snowy owls, and rough—legged buzzards, also rise in numbers, lagging a year or two behind the lemmings.
- Eventually conditions return to normal, the predators get the upper hand, lemming numbers fall—and the cycle starts again.

SCAVENGING

All animals and plants die eventually, although some are very long-lived. What happens to all the dead bodies, and to all the droppings the animals produced during their lives, and to the leaves that drop from the trees? They become food for scavengers—animals and other living things that specialize in eating the corpses of other creatures.

THE VARIETY OF SCAVENGERS

Scavengers vary from majestic vultures and eagles, to beetles and crabs, to worms and maggots. They play a specialized role in ecology as detritivores, feeding on detritus—the mix of dead animals and plants and their products.

LIFE FROM DEATH
Vultures on the African plains jostle for position on the carcass of an antelope. They strip the bones bare. Hyenas and jackals also join the feast.

Microscopic organisms such as bacteria and fungi finish the job, gradually rotting and recycling even bones, teeth, horns, claws, and other leftover hard parts, back into the soil to provide nutrients and minerals for new plant growth.

Common scavengers:

- Gulls such as herring and black-backed gulls are renowned scavengers, and can often be seen circling over garbage dumps where they will seek out any edible remains. They also feed on any dead animals they may come across.
- Crows often help to keep our roads clean by eating the remains of small mammals killed by traffic.
- Some flies lay their eggs on the bodies of dead animals. When the eggs hatch, the maggots feed on the dead flesh, helping to break it down.
- Other insects involved in waste disposal are the sexton beetles, which burrow beneath a corpse. They lay their eggs in a chamber dug from the soil nearby. When the larvae hatch, they feed on the decaying flesh.
- Many kinds of dung beetle feed on animal dung and lay their eggs in it.

CAMOUFLAGE

Many animals have colors and patterns to their fur, skin, or plumage, which make the animals difficult to see against the background of their normal habitat. For example, most female ducks are dull brown and mottled or barred, to blend in well as they sit tight on their nests in the reeds and bankside plants. Most owls have a barklike plumage so that it is hard to see them as they sit still during the day, tight against a tree trunk.

SHAPE AND COLOR

One of the finest examples of elaborate camouflage is shown by a tropical seahorse, the leafy sea dragon, from Australia. It is not only coloured yellow-green like some seaweeds, its body even mimics the feathery fronds of the weeds.

Similarly, it is not just the color, but also the shapes of stick and leaf insects that make their camouflage so perfect. They even sway with the breeze, using simple behavior to mimic their surroundings.

FADING LIGHT
In bright daylight, animals such as gazelles on the African plains may look clear and obvious. But in twilight, when many animals hunt, they blend into the background of shadowy grasses and bushes.

218

COLOR CHANGE
Chameleons are famous for their ability to blend into their backgrounds, and also for the fact that they can change color quite quickly if moved to a darker or lighter location.

Camouflage champions:

• Tree frogs may be easy to spot against the clear glass of an aquarium. But in nature, their green skin makes them almost impossible to see when hiding in foliage. Using color receptors in their eyes, they choose leaves specially to match their skin.

• Mammals and birds that stay year-round in the Arctic, such as the arctic fox and ptarmigan, change their coat or plumage to match the altering hues of their habitat. They grow a white winter coat, then molt to brown for the summer.

• One of the fastest color changes in the animal world is shown by the cuttlefish. It can turn from pale pink to almost inky black in less than one tenth of a second.

219

DEFENSE

Instead of hiding, another method of survival involves some animals making themselves unattractive to a potential predator. Animals, such as tortoises, tuck themselves away inside their hard shells so that the attacker cannot get at their soft flesh inside. Armadillos roll up and gain protection from their "armor" plating. Hedgehogs do the same and become spiny balls.

STARTLING COLORS

Some moths startle or bewilder their attacker by suddenly flashing their wings open to reveal bright "eye-spots." Frilled or bearded lizards spread out their bright, rufflike collars to scare an aggressor in the same way. This is called startle defense.

Many reptiles, especially lizards, can actually drop off the end of the tail, leaving the

DEADLY SKIN
The arrow-poison frog's skin is so toxic that just a smear from its oily surface can cause a small predator to suffocate and die.

Some animals, such as certain caterpillars and sea slugs, feed on poisonous food. They are unaffected but take the toxins into their own bodies, to deter attack from their own predators.

predator with the still wriggling piece. But the most important part of the lizard escapes and its tail slowly regrows.

WARNING COLORS

Many animals have evolved bright warning colors. These are aimed at relatively intelligent predators (such as mammals or birds) who learn to avoid these poisonous or unpleasant-tasting creatures by sight.

AWFUL SMELL

The skunk's vivid black and white colors warn that the animal can spray a foul liquid.

Mimicry:

- Some harmless creatures protect themselves by looking like, or mimicking, a similar but much more dangerous species.

For example, the hoverfly has a black and yellow striped body, like a wasp. But it has no sting and is harmless—unlike the wasp.

- The creature that is mimicked is called the model.

- Mimicry is especially common among insects such as butterflies, flies and beetles such as ladybugs, and also some types of colorful snakes.

- In another type of mimicry, very different animals have the same basic warning colors, so that predators come to associate these with unpleasant encounters. Common examples are yellow and black spots or stripes, and red and black patterns.

221

TERRITORIAL BEHAVIOR

The bird singing loudly from a garden tree in spring is not full of joy. It is involved in the serious business of territorial behavior. Using song, it is staking out a breeding ground and deterring invasion by another same-sex member of its species, while at the same time attracting a likely mate.

Some birds, such as the far-from-peaceful European robin, get into fights on the borders of their patch.

GARDEN TERRITORY
As birds battle for territories in our parks and gardens, they may use features such as fences and hedges as border markers, just as the human inhabitants do.

They may even battle to the death. A good territory, with its supplies of food, is a prize possession if the bird is to be successful in rearing its brood.

SCENT MARKING

Many mammals mark out the boundaries of their territories using scent. They leave smelly piles of dung or

LIVING SOLO
Birds such as the belted kingfisher have their own territories outside the breeding season.

SPRAINTS
Otters leave small piles of dung, known as spraints, on the riverbank to indicate their territory.

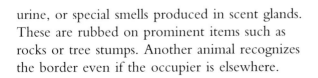

urine, or special smells produced in scent glands. These are rubbed on prominent items such as rocks or tree stumps. Another animal recognizes the border even if the occupier is elsewhere.

Unusual territorial animals:

- Some insects show territorial behavior. Large dragonflies can often be seen patrolling backward and forward around their chosen bit of pond or river. When a rival enters the air space, the owner promptly attacks it and tries to drive it away.
- Some butterflies are also aggressive to neighbors. They defend a patch of hedgerow or woodland and flap wildly at any invaders.
- Sea anemones jostle for space on seashore rocks. If one gets too near, its neighbor shuffles over and shoves it along, out of the way.
- Limpets graze algae on shore rocks, and then return to their home base, a small hollow in the center of their territory. If another limpet intrudes, the occupier slowly but surely pushes it away.

223

ANIMAL SOCIETIES

 "Society" is not normally a word associated with animals. In fact, many animals live in groups, sometimes in huge colonies. However, they may simply be together for shelter or to eat a local food source.

In a true society, members interact and communicate with each other. The main types of social animals are insects, birds, and mammals.

The social insects, such as ants, bees, wasps, and termites, live in colonies sometimes numbering millions of individuals. They all work together to maintain their safety and produce more offspring. In these sorts of societies, the individuals are all very closely related, and most of them are a special worker "caste," unable to breed themselves.

Some birds nest in colonies, or

SOCIABLE KILLERS
Killer whales are the largest members of the dolphin group. Also known as orcas, they live in groups known as pods, which usually contain related individuals. They communicate using clicks, squeals, and other sounds. Killer whales spread out to herd fish or seals into a bay, and then attack as a team. Each member recognizes the others as individuals, by their sounds and also by the details of their black and white body markings.

close together. But usually each pair of birds is separate, and gains protection from enemies by clumping together with other pairs, as in large colonies of seabirds.

MAMMALS

Many mammals, notably monkeys and apes, but also dogs such as wolves, show more obvious social behavior. The only social cats are lions. In their group or pride, the females work as a team to do most of the hunting, and bring back the kills for others to feed on. The males mark the pride's territory and repel invaders from other prides.

NOISY MONKEYS
Baboons live in troops
and have many different
sounds and signs to help
communication. There
are various alarm calls
warning about predators,
and about other baboons
invading their territory.
A new food find is also
signaled by another type
of call.

Sociable like us?

- Chimpanzees show the most complex social behavior in the animal world.
- Their behavior is very similar to another sociable ape—the human—so we can understand and appreciate it more readily.
- Chimps communicate with other members of their group using a huge range of facial expressions, sounds, gestures with their hands, and body postures.
- We must be careful in our interpretations of these actions. They may look like human behavior, but often they have very different meanings for another chimp. For example, when a chimp bares its teeth in a grin, it is not showing happiness like us, but fear and possible biting in self-defense.

225

COURTSHIP

Before reproducing, many kinds of animals show elaborate courtship behavior. This has evolved to make sure that partners are from the same species, of different sexes, healthy and suitable for each other, before they put their energies into producing eggs or offspring.

The best known courtship patterns are seen among birds. The males have bright, impressive plumage, loud songs and even energetic "dances" which make them particularly obvious at breeding time.

MALES ON SHOW

The male quetzal of the South American tropical forests displays its bright green plumage and sports its long tail when courting a female.

The cock-of-the-rock, another South American species, meets other males in traditional places, called leks. Their courtship displays involve strutting and dancing, and showing off their bright orange plumage to attract a female.

The spectacular eyed tail fan of the male peacock is used solely for courtship display, to impress a mate.

The Great Frigatebird male inflates his scarlet throat-sac, spreads his wings and makes throaty cries to attract a mate. A female will respond by nibbling his feathers and rubbing her head against his pouch.

226

RUTTING TIME

Many mammals, such as gazelles, deer, and antelopes, spend a great deal of energy in courtship. This often involves stylized battles between rival males, as in rutting contests. The contestants clash heads, and push and shove to show their strength and endurance. Serious injury is rare, however.

BEACH BATTLES
Male seals battle not for a female, but for a small patch of beach as a breeding territory. Only when a male has secured this territory, is he able to mate.

More courting couples:

In the vast majority of animal species, it is the male who has bright colors, loud calls or special actions. The female is usually less prominent.

- The loud croaking and chirruping of frogs is part of their courtship. So are the ballet-like dances of the Siamese fighting fish.

- Insects such as butterflies and mayflies flit together in nuptial flight.

- The male field cricket chirps at his burrow entrance to alert nearby females.

- Sometimes females are the more active partners. Female glow worms (actually a type of beetle) are wingless. They flash their lights to attract the winged males.

227

EGGS AND NESTS

The hard-shelled eggs that birds lay are a splendid adaptation to life on land, even in dry and desert conditions. However, eggs are vulnerable to certain predators and must therefore be protected—either by the adult birds, or by being concealed in a nest.

Many eggs are laid in the open. But they are patterned to match their background, with irregular blotches or squiggles.

Various seabirds find safety in numbers, and also by choosing nest sites that are difficult for predators to reach, such as sea cliffs or isolated islands.

INCUBATION

Birds must incubate their eggs, which means keeping them warm enough in cold conditions so that the embryos (babies) can develop inside. In hot climates, the parent birds often have the

AWKWARD ACCESS

Weaver birds, such as the red-headed weaver, build their nests from stems, leaves, moss, grass, and thorns woven together into a strong structure. The nest is suspended from a branch tip and its opening faces down, so it is difficult for predators to enter.

228

Village weaver birds build their nests woven together into a colonial structure, which may be 16 feet (5 m) across. If a predator tries to steal the eggs, all members of the group squawk alarm calls and flap nearby. These noisy birds live in Africa, in a variety of habitats including farmland and parks.

opposite problem. They must prevent the eggs overheating, by shielding them from the direct sun, or even by wetting them from their own breast feathers.

Incubation times vary from species to species, from about 10 days in small songbirds to as long as 80 days in albatrosses.

HELPLESS OR ABLE?

The young of many birds, such as thrushes and other songbirds, are naked and helpless for the first couple of weeks. They must be tended and fed by their parents.

In others, for example game birds, waterfowl, and shorebirds, the chicks have feathers and can run and even fend for themselves soon after hatching.

Egg and nest records:

- The largest egg is that of the ostrich. It can be 8 inches (20 cm) long and weigh up to 4 pounds (1.9 kg).
- The largest egg relative to body size is laid by the kiwi. It may be almost one quarter of the mother's body weight.
- The smallest eggs are laid by hummingbirds, and may be less than 0.4 inch (10 mm) in length.
- Hummingbirds also make tiny nests—thimble-shaped cups woven partly from spider webs.
- Some birds, such as the fairy tern, make no nest at all. This tropical seabird balances its egg on a bare tree branch, hoping for calm weather.
- The largest single nests are made by eagles over many years. They can be 10 feet (3 m) across.

SAVING WILDLIFE

As the pressures on the natural world increase, especially due to our growing human populations, the wildlife of the planet becomes ever more threatened. Such pressures include pollution of the environment from industry, traffic, and agricultural chemicals, and the destruction of forests, grassland, wetlands, and other habitats to create farmland, housing, and roads.

There are still many wild places left on earth where plants and animals flourish almost unhindered, such as the open Siberian tundra and boreal forests. But the richest places for wildlife are also some of the most threatened—particularly the moist lowland tropical forests, which contain far more species than anywhere else, but which disappear at an alarming rate. In highly populated temperate

regions, such as western Europe, there are few natural habitats left, and any remnants are therefore very precious.

FAMILIAR BUT ALMOST GONE

Many animals familiar to us from zoos and wildlife parks are under real threat in their natural homes. Examples are black rhinos, the great apes (orangutans, chimpanzees, and gorillas), tigers, and that famous symbol of conservation, the giant panda. Without adequate protection of their habitats, these species (and many others) are almost certainly doomed to extinction.

EXTINCTION OF WILDLIFE

Throughout the long history of life on Earth, species have become extinct. It is a natural part of evolution by natural selection. Scientists estimate that more than 99 percent of all the species of animals that have ever lived have become extinct. The main evidence for them is fossils. They include a host of invertebrates, such as trilobites, as well as larger creatures such as the dinosaurs.

SYMBOL OF EXTINCTION
The dodo was a turkey-sized flightless pigeon from the Indian Ocean island of Mauritius. Inoffensive, and a useful source of fresh meat for passing sailors, it became extinct at the end of the seventeenth century.

FASTER AND FASTER
In recent decades, the activities of people have caused an increasing number of animal and plant extinctions, and much of our wildlife is today threatened as never before. Animals which evolved on isolated oceanic islands have been particularly vulnerable. Most live nowhere else, and cannot be relocated because they are so closely adapted to their unique island habitat. Many have been driven to extinction by introduced predators such as cats and rats. Others suffer at the development of tropical "paradise islands," as the natural vegetation is cleared for the tourist and leisure business.

EXTINCT OR NOT?

When is an animal truly extinct? The Tasmanian wolf, also called the Tasmanian tiger or thylacine, was a doglike mammal from Tasmania, south of mainland Australia. The last known individual died as a captive in the zoo at Hobart, the Tasmanian capital, in 1936.

The species is now presumed extinct. But occasional claimed sightings persist. Some experts believe the thylacine lurks in the dense Tasmanian forests and, like the coelacanth of African waters, waits to be rediscovered.

NEVER SAY DIE?

The coelacanth is an unusual fish with a fleshy lobe at the base of each fin. Its fossils are known from the time of the dinosaurs, but then it seemed to die out—until 1938, when a coelacanth was caught off the east coast of Africa. The fish was known to local fishermen, but scientists were amazed. Since then several more have been caught and studied.

Examples of some animal extinctions:

Moa

The moas of New Zealand were flightless birds taller than a human, resembling giant kiwis. They were hunted to extinction by the early human settlers, finally dying out around 1800.

Great auk

This flightless relative of guillemots went extinct in 1844 when the last known bird was killed near Iceland.

Quagga

This zebra-relative from South Africa was hunted to extinction in the 1880s.

Passenger pigeon

This pigeon used to occur in huge numbers in North America, turning the sky black for hours in its gigantic flocks. But it was hunted to extinction in the wild by about 1900.

233

HABITAT DESTRUCTION

 The destruction of habitats is the greatest of all threats to wildlife, be they rich tropical forests, mangrove swamps, coral reefs, or your own local grassland or wood. Most wild plants and animals are so closely adapted to their own particular habitat that they become rare or endangered if this is damaged or removed.

MORE HABITAT

The paradise whydah is a bird of dry open country in Africa. As farmland spreads at the expense of forests, its habitat is actually becoming larger. However, the male is hunted for its spectacular tail feathers.

MOST AT RISK

Globally, the most worrying losses of habitat are the tropical rain forests, because these contain by far the largest number of species. Although large areas of tropical forest still survive, they are still being lost at an alarming rate—areas the size of small countries each year.

Coral reefs, another rich habitat, are threatened by overfishing and shell collection. But perhaps a greater threat is the mud and silt from land erosion, which enters the shallow-sea coral area from nearby rivers, and kills the live corals.

LOGGED AND GONE

After an area of tropical forest is logged and cleared, it takes at least a century to regrow.

More threatened habitats:

Wetlands

- Marshes, swamps, and other wetlands everywhere are very vulnerable to pollution of the waterways that sustain them. Their plants and animals are easily poisoned.
- Many wetlands are viewed more as "wastelands." They are drained to provide more and more land for growing crops.

Mangrove swamps

- These coastal subtropical and tropical areas of mangrove trees are under threat in many places, especially in Southeast Asia, where they are often converted to rice paddies.
- But when the mangroves have been removed, the coast becomes much more vulnerable to erosion by storms.

HUNTING

People have hunted animals and collected wild plants for thousands of years. In the early days of human evolution, it was necessary to survive. Nowadays, however, hunting continues mostly as a sport, or in more sinister fashion as illegal poaching for profit. Beautiful shells and some kinds of wild plants, including cacti, are collected as well.

HUNTING AT SEA

The large whales were hunted almost to the point of no return for their meat, oil, and fat. In the 1980s, most countries halted this activity and whale populations now show signs of recovery. Dolphins, smaller cousins of the whales, suffer from being snared in fishing nets and many die accidentally by drowning.

SUCCESS STORY?

Begun in the 1960s, Operation Tiger planned to save these big cats. But in the 1980s, it was discovered that the numbers of tigers had been deliberately recorded as too high, giving a false impression of success.

NOT SO COMMON
Common dolphins have become rare in some areas, where they suffer pollution or get trapped in nets.

THE IVORY TRADE

Hunting has had a major impact on large mammals, especially on the open plains and savannas of Africa where rhinoceroses, elephants, and others are easily spotted, and shot. The high price of ivory has fueled the illegal killing of elephants, which have become rare in many areas.

CLUBBED TO DEATH
A film of baby seals being clubbed to death for their fur offended many people around the world. Most seals are now safe, but many other animals are still treated in this way.

Continued exploitation:

- Live monkeys and apes, and tropical birds such as parrots, are trapped and exported, usually illegally. Many of these die in transit.

- Reptiles, such as crocodiles and alligators are trapped, and their skins exported, again illegally, for making fashion items such as handbags.

- Big cats such as tigers and leopards are still also killed for fashion, their skins finding their way eventually into hats and coats.

- Smaller and lesser-known cats are also under threat. Their illegally obtained furs are less easy to identify and trace.

- Rhinoceroses are killed for their horns. Some people believe that powdered rhino horn has medicinal powers.

237

POLLUTION

 Pollution comes from many different sources: chemicals draining from farmland, factories, or sewage outflows; fumes pouring from vehicles, factories, and power stations; and events such as leaks from oil pipelines or tankers at sea.

ACID RAIN
Every year pollution kills countless numbers of wild animals and plants. It

CHEMICALS IN THE WATER
The valuable metal gold is sometimes extracted from rivers using the poisonous metal mercury. The mercury washes downstream with devastating results, killing fish and plants.

can even affect nature reserves and other wildlife refuges. A particular problem is caused by acid rain. Rain becomes unnaturally acid when it absorbs sulphur and nitrogen, which

RISING SEAS
Global warming was detected in the late 1990s by climate experts. As the polar ice caps melt, sea levels could flood vast areas of low-lying land. This would destroy valuable wildlife habitats such as coastal marshes and mangrove swamps—and also hundreds of ports and coastal towns.

are put into the air mainly from the burning of coal, gas, and oil, used as fuel for cars, factories, houses, and power stations. This acid rain reduces the fertility of the soil and causes damages trees and water life.

Acid rain usually falls hundreds of miles away from the polluting area. It is an international problem and has already damaged huge tracts of forests in northern North America, Europe, and northern Asia.

Pollution high in the sky:

• In recent years scientists have found that several chemicals have damaged the upper atmosphere, with far-reaching effects.

• The upper atmosphere contains a layer with high levels of a gas called ozone. When this gas forms at low levels it can cause illnesses such as asthma.

• But the natural ozone layer of the upper atmosphere protects the Earth from the harmful ultraviolet (UV) radiation from the Sun.

• Scientists have found that there is a thinning or hole in the normally complete ozone layer. This is growing, damaged by several different chemical pollutants.

• International agreements have limited the production of some ozone-destroying chemicals. But the problem will persist for many years to come.

PESTS

When species are brought or introduced to new areas, either deliberately or by accident, they may upset the delicate ecological balance. This can have devastating results. Introduced species can quickly become animal pests or plant weeds, outcompeting the native animals or plants and causing local extinctions. Such pests can then be extremely difficult to eradicate or control.

Introduced predators include cats, rats, weasels, pigs, and mongooses. They have especially severe effects on small islands where the ecology is already fragile. Goats, deer, pigs, and rabbits are commonly introduced and

RATS AND RABBITS
Rats are supreme survivors. They thrive on the same foods we do, whether these are growing in the fields, stored after harvest, or thrown away as leftovers. Rabbits hide in their burrows, eat our crops or grazing grassland, and breed so rapidly that they have spread to every continent except Antarctica.

BIOLOGICAL CONTROL

The prickly pear cactus, originally from North America, was introduced to Australia, where it soon become a weed. However, it has been partly controlled by another introduced species, a moth from Argentina, whose caterpillars eat the cactus.

have had a dramatic impact on islands with no native browsing herbivores, such as in New Zealand and Hawaii.

OUT OF CONTROL

Some weed or pest species, such as rats and mice, and plants such as plantains and pineapple weed, have managed to spread themselves around the globe without special help from people. Despite precautions, they manage to evade our controls.

As genetically modified crops and farm animals begin to appear, will they also escape and become established in the wild, adding to the chaos?

An infamous weed—the water hyacinth:

- One of the most notorious of all introduced weeds, this is a floating lilylike plant from tropical America.
- Its large leaves and attractive purple flowers have choked waterways in many countries, becoming a very serious nuisance.
- Water hyacinth has spread to the southern USA, Africa, Australia, India, Sri Lanka, Malaya, and even to Tahiti and the Solomon Islands. In many places it has altered the ecology and impeded boat traffic, especially clogging boat propellers.
- In just three years it choked some 900 miles (1,500 km) of the Zaire River.
- As it dies and rots, water hyacinth uses up valuable dissolved oxygen in the water, killing animals such as fish.

241

SAVING SPECIES

The most effective and natural way to save threatened species is to ensure that they have sufficient and suitable places to live—their natural habitats. It is also vital to keep human pressures, such as hunting, to a minimum.

However, this is not always possible, and emergency measures may be taken to save species on the brink of extinction.

Some species have been taken into captivity and bred there to build up their populations, before being returned to the wild. Success has been achieved with the European bison, the American buffalo (bison), and also with the Arabian oryx. Przewalski's horse is another candidate. Herds of this domestic horse ancestor have been established with the hope of reintroduction to their native habitat, the Asian steppes.

SUMATRAN RHINO
Critically endangered, this species probably breeds too slowly to be saved in zoo programs.

MEDITERRANEAN MONK SEAL
This seal is very nervous of human presence, and mothers may lose their unborn babies if disturbed. However, the seal's range is becoming more crowded with holiday centers, boats, and scuba divers.

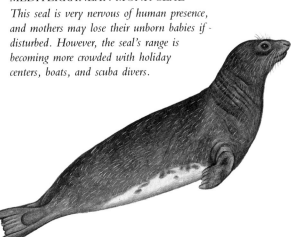

Several zoos now specialize in breeding rare species, with the hope of reestablishing them in the wild. However, zoo animals may become too dependent upon people.

GLOVED HAND
The California condor, one of the world's largest flying birds, became extinct in the wild in the late 1980s. A breeding program has kept it going in captivity. The chicks are fed by hand, using a special glove shaped like an adult condor's head. Hopefully, the young birds will not become familiar with their keepers.

Just a few very rare animals:

Black-footed ferret
USA
- Threats are habitat loss, disease, hunting (next page)

Blue whale
All oceans
- Main threat was hunting

Aye-aye
Madagascar
- Main threats are deforestation, hunting

Golden lion tamarin
Brazil
- Main threat is deforestation

Orangutan
Borneo, Sumatra
- Main threats are habitat loss, capture

Philippine eagle
Philippines
- Main threats are habitat loss, hunting

243

PARKS AND SANCTUARIES

Wildlife reserves, national parks, and sanctuaries of various kinds have been established throughout the world, to try and preserve as much of our wildlife heritage as possible. Some seek to exclude people, while others try and involve local people, especially where traditional use of the habitat combines with conservation.

ECO TOURISM

In most cases the reserves may be visited by naturalists and tourists. But there are usually restrictions to protect the wildlife. Eco tourism, where income from visitors helps conservation projects, is a fast-growing business.

BLACK-FOOTED FERRET
Prairie dogs are the main food of this agile carnivore. But prairie dogs are also pests, and often poisoned. The poison accumulates in the ferret's body and causes illness and death.

PLATYPUS
River pollution has affected the worms, shellfish, and other foods of this unique Australian monotreme (egg-laying mammal).

The Everglades, a huge swamp at the tip of Florida, is a famous wildlife reserve. It is fed by fresh water seeping through it from a lake and river. The water forms pools, marshes, and meandering channels—it is one of the greatest wildlife sites in the world. The heart of this ecosystem is protected by the Everglades National Park, a Biosphere Reserve and World Heritage Site covering 1.4 million acres (566,000 ha).

EVERGLADES INHABITANTS

The Everglades is home to rare species such as the wood stork, Everglades kite, reddish egret, and the endangered Florida "panther," a local version of the puma (cougar or mountain lion).

SNAIL FOOD
The Everglades kite is a local form of the snail kite of Central and South America. It feeds almost entirely on one kind of water snail, extracting the flesh using its specially adapted narrow, curved bill.

Some notable wildlife parks, and marine sanctuaries:

Baikal, Russia
- The world's deepest lake, and over 25 million years old. It contains about 960 species of animal and 400 species of plants found nowhere else in the world.

Fjordland National Park, New Zealand
- Unspoiled forest and hills, and home to rare birds such as the takahe and kakapo.

Galapagos National Park, Ecuador
- Made famous by naturalist Charles Darwin, the island inhabitants include Darwin's finches, giant tortoises, and marine iguanas.

Great Barrier Reef, Australia
- This vast coastal and underwater reserve protects the world's greatest coral reefs.

THE FUTURE FOR WILDLIFE

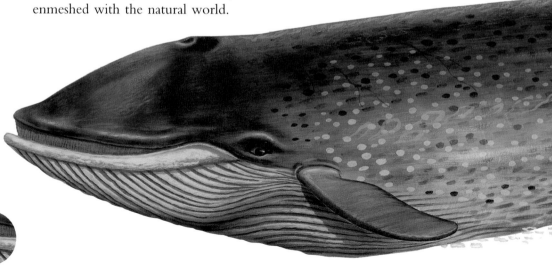

Traditional human societies developed in close association with the natural world. They devised ways of using its products, working with rather than against nature, in such a way that did not exhaust supplies. Those early people were generally in balance with nature.

Modern society is not so closely enmeshed with the natural world.

Most people have little contact with wild plants and animals. Much of our involvement with wildlife is now destructive. This is made more acute by the massive increase in the human population of our planet, with the need for ever more housing, farmland, and industry.

THE IMPORTANCE OF NATURE

We need to make people more aware of the fragile balance of nature, and its fundamental importance—not just to plants and animals, but to our own future.

For example, large areas of forest not only harbor fascinating wildlife. They also breathe back vital oxygen into the air, and affect rainfall and climate. In so many ways, the future of wildlife is our own future too.

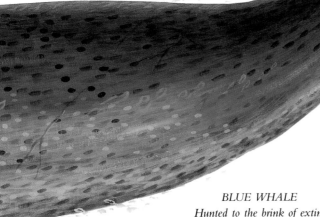

BLUE WHALE
Hunted to the brink of extinction,
the numbers of blue whales may slowly
be rising. These huge animals, biggest on the planet,
breed so slowly that their recovery will take centuries.

More notable wildlife parks and marine sanctuaries:

Serengeti National Park, Tanzania
• In East Africa, the savanna animals include cheetah, wildebeest, zebra, lion, leopard, elephants, gazelles, and black rhino.

Virunga National Park, Zaire
• The forest is best known for its mountain gorillas.

Wenchun Wolong Nature Reserve, China
• The main home of the giant panda.

Yellowstone National Park, USA
• Grizzly bears, wolves, cougar, bald eagle, and others roam this well-established park.

Yosemite National Park, USA
• Groves of giant redwoods, the world's largest trees, harbor animals such as black bear and bobcat.

247

GLOSSARY

angiosperm

A flowering plant—a plant that reproduces by flowers or blooms, which develop seeds. Angiosperms are the most complicated plants and include the vast majority of common flowers, herbs, grasses, bushes, and trees (except for conifers).

arthropod

An invertebrate animal with hard-cased jointed limbs or legs, rather than flexible arms or tentacles. Arthropods include insects, crabs and other crustaceans, spiders and other arachnids, centipedes, and millipedes.

broad-leaved

A tree with broad, flat leaves, as opposed to the needle-shaped or scaly leaves possessed by conifer trees. Most trees in the flowering plant group (angiosperms) are broad-leaved. Most are also deciduous, losing all their leaves at a certain season (such as winter), rather than evergreen.

carnivore

A carnivore is an animal that eats the meat or flesh of other animals.

cold-blooded

An animal that cannot produce heat within its body to keep it warm. So its body temperature fluctuates, being warm in hot conditions and cool in cold conditions. All main animals groups except for mammals and birds are the cold-blooded. A more scientific term is *exothermic,* "heat from outside."

deciduous
A tree, bush, or other plant that loses all of its leaves at a certain season (such as winter or the dry season), as opposed to an evergreen plant that always has some leaves.

detritivore
A living thing that gets its food from detritus—dying or dead plants, animals, and other organisms and their products, such as dung and rotting wood.

ecology
The study of plants, animals, and other living things, and how they interact and survive in their nonliving surroundings.

evergreen
A tree that does not lose all of its leaves at a certain season (such as winter), like a deciduous tree. It loses a few leaves regularly all the time and soon replaces them, so it always has some greenery.

evolution
The gradual change in living things over time, usually very long periods of time—thousand and millions of years. This happens in response to the changing conditions and environment. Some species of living things adapt to the new conditions, by the process of natural selection, and evolve into new species. Others die out or become extinct.

flower
The part of a flowering plant or angiosperm specialized for reproduction, to produce seeds.

food chain
The sequence of events where a plant is eaten by an animal (herbivore), then the animal is eaten by another animal (carnivore), and so on. Most plants and animals are involved in more than one food chain, and food chains usually link together to form much more complex food webs.

habitat

A certain type of area or natural place, such as a woodland, a sandy seashore, a desert, or a city park. Each kind of plant, animal, or other living thing is suited, or adapted, to surviving in a particular habitat.

herbivore

An animal that eats plant food, such as leaves, fruits, seeds, shoots, and roots.

invertebrate

An animal without a backbone (spinal or vertebral column) is called an invertebrate. The vast majority of animals belong to this group, both in numbers of species and actual numbers of animals.

microbe

Any microscopic living thing, such as a bacterium, protist or moneran, a spore, or pollen grain, or some of the smallest fungi, plants, and animals.

omnivore

An animal that eats a variety of plant and animal food.

organism

A living thing—plant, animal, fungus, moneran, or protist.

pollen grain

Tint dustlike grains produced by the male flower of a plant. They contain the male sex cells, and convey them to the female cells of a flower of the same species, for reproduction.

pollination

The transfer of pollen grains from the male parts of a flower to the female parts, usually of another flower, for reproduction. This can be done by wind, water, or animals such as insects, bats, and birds.

seed

In plants, the part produced by reproduction that contains a tiny embryo of "baby" plant, ready to grow, plus a food store for this young plant. Fruits and nuts contain seeds. See also spore.

species

The basic grouping of living things. All members of a species can breed together to produce more of their kind. But they cannot breed with members of other species, or if they do, the offspring are not fertile (able to breed themselves). So all tigers are one species. All lions are another species.

spore

One or a few microscopic cells that will grow into a new individual plant or other living things. Fungi and simple plants reproduce by spores, which mostly look like tiny grains of dust. More complicated plants, such as conifers and angiosperms (flowering plants), reproduce by means of seeds.

vertebrae

The individual bones or parts of the spinal column, or vertebral column, or a vertebrate animal. The vertebrae of some vertebrates are not bone, but cartilage, as in sharks.

vertebrate

An animal with a backbone (spinal or vertebral column). The main groups of vertebrates are fish, amphibians, reptiles, birds, and mammals.

warm-blooded

An animal that produces heat within its body to keep it warm, so its body stays at a constant temperature, even in cold conditions. Mammals and birds are the only two main groups of warm-blooded animals. A more scientific term for this is *endothermic*, "heat from inside."

INDEX